# Survivor

This is the story of a 1929 Rolls-Royce 20/25. It chronicles the birth, the history, the mechanical puzzles and the solutions found, in bringing this car back to good and reliable condition. Plus data and servicing information, updated to take advantage of the availability of modern high tech oils and materials. This book is not intended to be a workshop manual and nor does it claim to be comprehensive. It is intended to help the average owner, care for and improve, their own special car.
"The quality will remain, long after the price is forgotten", Henry Royce once sagely observed.
That's still true today.

Charles Vyse July 2013
The current custodian of GXO71

Further copies of this book are available from: http://www.lulu.com/shop (listed under 'engineering')

# Contents

*All technical information is given in good faith, but you follow it at your own risk. Acknowledgements are made for help received over the years, from Will & Rob at Fiennes, the team at West Hoathly Garage, Tom Clarke, Phil Cordery, David Else, Dan Suskin and Stephe Boddice. Also Dean Ash, Simon Slaffer and Keith Whitehead for some images. Any errors in this book are entirely my own work.*

# Part One
# *In the beginning...*

It was the summer of 1929. Alf Pearce, an apprentice at the Rolls-Royce Derby factory, had been given a tedious but important job. He was crouched over the skeletal beginning of a new 20 horsepower model chassis. Mounted on trestles either side of him, were the sturdy 'U' section side rails for the chassis, fresh from the fabrication shop. On a bench alongside sat the chassis cross members, waiting to be fitted. A bucket on the floor was filled with newly machined tapered bolts.

Alf's job was to place each cross member into position across the rails and then selectively fit a tapered bolt into the tapered holes. These high tensile bolts, together with their nuts and locking washers, held the completed chassis together. He was not allowed to bang the bolts fully home; that would be the job of the journeyman craftsman he was apprenticed to. Alf was required to tap each bolt head lightly with a mallet and judge if it was tight enough in the taper, to be rigid when finally hammered home. But not so tight, that the hammering force required would distort the bolt head. He tried bolt after bolt in each hole, until he found the one that fitted 'just so'. When his work was checked, if a bolt was judged to be a poor fit in the tapered hole, then in all probability it would be removed and thrown at him. The journeyman who instructed Alf, explained why correct fitting was so important. Pa Royce (as the apprentice's referred to him)

had found that rivets came loose over time as the chassis flexed, when the car was being driven. As would ordinary bolts. But a tapered bolt hammered into a precisely undersize tapered hole and *then* correctly tightened down, would stay tight. Now that was a proper job. Which is why each bolt had to be selectively fitted and the nut tightened, using the correct spanner.

Rolls-Royce designed the spanners themselves. Each spanner was exactly the right length, that Henry Royce calculated would apply the correct torque to the nut, when heaved by an average strength man. Though the journeyman fitters were so experienced, that they could judge torque being applied just by feel. Henry Royce, who was the engineering genius behind Rolls-Royce, had just one standard. Absolute perfection. And if you didn't measure up, then you were sacked. It was as simple as that. And which was why each chassis took three months to complete, before the chassis could be sent to a coach builder, for a body to be constructed. This particular chassis that Alf was working on, had yet to find a customer. Chassis GXO71 when complete, was marked on the Chassis Card to be put into 'stock'. Hopefully, it would not be too long before it was sold. Hopefully, because on the economic front in Great Britain, things were not looking at all rosy.

In America, there was also an intangible feeling in the air that economically there was something amiss. Famously this was put into words by the financier J D Rockefeller, who remarked that he knew it was time to sell all his Wall Street stock, when he heard a pavement boot-shine boy, recommending a sure fire stock to buy, to the guy whose shoes he was shining. But there was no signs of gloom, economic or otherwise, on the penthouse floor of a very expensive Manhattan apartment. The owner, Guy Mitchell, had just finished a fond letter to his very special British friend, Walter Chapman. Walter lived in Boxmoor, Hertfordshire. His relationship with Guy had begun, when he was appointed as Guy's British Business Agent. But over time, their friendship had moved onto another level altogether. Both men were in fact gay and in a relationship. However this was a long distance affaire,

## IN PURSUIT OF PERFECTION

*Henry Royce trusted very few manufacturers. Consequently Rolls-Royce made almost everything themselves that went into their cars. The battery, sparking plugs, wheels, tyres & electrical cable were about the only exceptions.*

*Take the basic nut & bolt for instance. Most motor manufacturer's bought these off the shelf from specialised companies. Henry made his own. Most fasteners as they are known, were made from mild steel and had hexagon heads. The Rolls-Royce bolts however, were made from 3% nickel steel; harder than the mild steel used for the nuts.*

*This meant that if over zealous tightening resulted in a stripped thread, it was the cheaper and easier to replace nut that suffered and not the bolt. Additionally, the bolts were made with a square, rather than the common hexagon head. This meant that the bolt could be held from turning, by the use of a step or spigot on the component.*

*"Attention to detail" was the watchword in the Derby factory.*

as Guy preferred to live in the US and Walter in the UK. Guy usually spent a month or so each year visiting the UK, crossing the Atlantic at the beginning of every August on one of the Cunard luxury ocean liners. Guy's father had founded a Bank in Illinois, USA and Guy was extremely wealthy by any standards. He rented an apartment in London's West End on an all year round basis, even though he used the apartment for very few weeks in any given year. He and Walter would spend time together when Guy was in the UK. Although it all had to be kept very discreet, as homosexuality was still a criminal offence in the UK in those days.

On Guy's trip to the UK in August 1929, he and Walter took a holiday together to Scotland, traveling north by steam train. It was a splendid holiday, where Guy had first tentatively started to trace his own Scottish ancestry. He wanted to spend much more time in Scotland and to really get to see the Highlands. To do this he would need a car. There were two problems. Firstly he couldn't drive and secondly, he didn't have a car. So he decided to buy one and then employ a part time chauffeur. He kept an apartment in the UK; so why not a car as well? Asking the more knowledgeable members of the various clubs that he belonged to, he discovered that the 'best' car in the world was generally considered to be a Rolls-Royce. And as Guy always had the best of everything, that was the car he decided to buy. Somewhat brashly, he wired Rolls-Royce long distance to place an order. To his surprise, he discovered that buying a Rolls-Royce was not that simple. Did he want a limousine? A Brougham perhaps? Or an open tourer? Maybe a saloon? With division or without? Owner driven or did sir have a chauffeur? And crucially, who was sir's preferred coach builder?

To Guy, these were a lot of damn

> **"Whatever is rightly done,
> however humble,
> is noble."**
> *Sir Henry Royce 1863 - 1933*

fool questions and delivered in that ever so slightly superior way, that came naturally to the British. Which he didn't care for. Buying a car in that goddam country he thought, involved the same sort of ritual baloney he got from his London tailor, when being fitted for a suit. Guy had more important things to think about. So he wrote

to Walter at his little house in Boxmoor, Hertfordshire and asked him to buy a Rolls-Royce on Guy's behalf.

Walter loved motor cars, though truth be told, he was an indifferent driver and had never owned a car himself. He knew Fred Snoxall, the proprietor of Snoxalls Garage in Boxmoor where he

> **"Strive for perfection in everything you do. Take the best that exists and make it better. When it does not exist, design it."**
> *Sir Henry Royce 1863 - 1933*

lived. Fred had been in and around motor cars all his life. Walter thought that if anyone knew the correct path to Rolls-Royce ownership, it would be Fred. So he went to see him. Once Fred realised that Walter's enquiry was serious and backed with serious money, he realised that if he secured the order, it would raise the profile of Snoxalls Garage. Fred had founded the garage just three years earlier in 1926 and up until Walter's enquiry, had dealt mainly in Austin and Morris motor cars. A Lanchester was the most exotic car he had sold to date. He made an appointment to see Major Len Cox at Rolls-Royce in Conduit Street, London. Major Cox was the Sales Director of the prestigious London headquarters. The meeting went well, but Major Cox told Fred Snoxall that there was a lot more to becoming a Rolls-Royce retailer, than an order for just one chassis. Major Cox also pointed out, that there was a three to six month delivery date for a new

Rolls-Royce chassis. However, larger approved retailers often ordered a chassis for stock and it was very possible, that Snoxalls Garage might acquire a chassis that way. But Snoxalls Garage would not be permitted to purchase a chassis direct from Rolls-Royce. Apart from any other consideration, Snoxalls Garage would have to employ Rolls-Royce trained mechanics, before any dealership could be considered. Major Cox gave Fred contact details of approved Rolls-Royce retailers and they parted amicably.

Within a week, Walter Chapman had written a long letter back to Guy in America. Walter proposed that the car most suited for long distance summer touring in Scotland, would be an all weather tourer. He went on to explain that this meant a fabric roof that could be lowered and side screens which could be completely removed. Guy replied that he would leave it all to Walter, but he would like to see an illustration of the proposed car. He also mentioned in his chatty letter, that he had been much taken with the

streamlined beauty of a car he had seen at Daytona Beach in the Spring. It had set a new flying mile world record at 231.45 mph. The car was called 'Golden Arrow' driven by Sir Henry Segrave. Guy had been told that the beautifully streamlined aluminium body had been constructed by a London coach builder. This letter prompted Walter to telephone the major Rolls-Royce approved coach builder's and to request

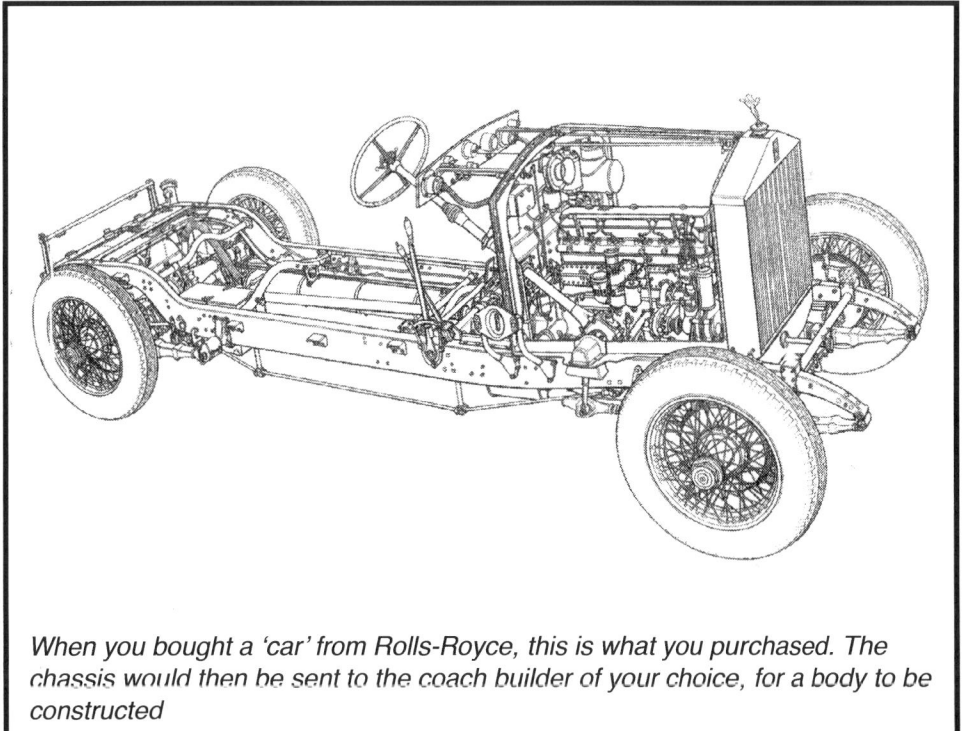

When you bought a 'car' from Rolls-Royce, this is what you purchased. The chassis would then be sent to the coach builder of your choice, for a body to be constructed

brochures. When these finally arrived in the post, Walter who was quite artistic himself, considered that many of the designs were clumsy, or ill proportioned, or both. Fred Snoxall advised him that Barker not only held the Royal Warrant, but in his opinion also built the best tourer bodies. But Walter did not care for the bulbous look of the front mudguards. He had also learnt that Barker bodies were heavier than most. And that Mr Henry Royce advocated the lightest body possible, in order not to sap power and dull the car's performance. So bearing Guy's letter in mind, he wrote to 'The Times' in London, requesting a back issue of their report on the land speed record of 11 March 1929. When he duly received this from 'The Times', he was surprised to learn, that Thrupp & Maberly who had built the aluminium 'Golden Arrow' record breaker, was also a Rolls-Royce approved coach builder. Walter had not contacted this firm before, because he believed that all they built were Limousines. The very next day he took a train to London and called on Thrupp & Maberly at 108 Cricklewood Lane. Had they ever built a tourer on the small Rolls-Royce chassis? Yes they had! Five to date. Even better, similar to their

rival Barker, Thrupp & Maberly had also recently been awarded the Royal Warrant. Walter knew this fact would impress Guy; like so many Americans, he just loved the British monarchy! However, the photographs of their tourer bodies that the Sales Director showed Walter, were similar to the Barker design, which he didn't like. With the help of the Sales Director, Walter searched through dozens of designs and pictures of cars from all over the world. He knew in his mind's eye what he wanted and eventually he found a visual reference. Then and there a draughtsman was called in and in no time at all, a very rough sketch was produced. This was more like it! The sketch showed an elegant design sculpted from aluminium over an ash frame. All four mudguards were 'helmet' style with a knife edge crease running longitudinally in line with the front side lights. Walter had mentioned to the draughtsman, that he intended to take the proposed car on long trips into the Scottish highlands. On the draughtsman's advice, a 'bridge' running

The Rolls-Royce assembly line in Derby where GXO71 was built. 20HP chassis are on the left and the bigger 40/50HP on the right.

transversely across the car behind the front seats was added. Scuttle shake when driving on bad roads was a universal problem. The 'bridge' would strengthen the body without adding a lot of weight. Thrupp & Maberly promised to add colour and work up the design, so that it could be sent to Guy in America for his approval.

When Guy eventually received the finished design proposal, he was thrilled. The lines of the car were beautiful. But Guy with the innate good taste of a man who collects Italian renaissance art, thought he could improve upon it. So he asked if his new car could be all polished aluminium, rather than painted. In 1929, unpainted cars were usually those headed for the tropics. Paint technology was still in its infancy and paint subjected to strong UV for any length of time soon degraded. When Thrupp & Maberly were later consulted by Walter, they suggested a compromise and produced another design sketch. Now the car had highly polished aluminium wings, front valance, bonnet top and scuttle. Further, each of the four doors had been given a rolled aluminium top, which matched the polished top of the bridge across the car. The doors, bonnet side panels and the body tub were all finished in a striking shade of dark blue. The rest of the car was highly polished aluminium. The leather upholstery and the fabric roof were both sand coloured. This was not a car you were ever going to not notice. Guy loved it!

Walter duly placed an order with Snoxalls Garage for a new Rolls-Royce 20HP chassis. He wanted the completed car delivered before the following Spring, so that he and Guy could take an early motoring holiday. Fred Snoxall in turn spoke to Major Len Cox. He was told that Rolls-Royce did have a 20HP chassis nearing completion and which was due to be delivered to Rootes Limited and was marked 'stock'. Rootes was a large retailer and by a fortunate coincidence happened to also own Thrupp & Maberly. Fred was able to negotiate a split commission deal with Rootes, on the understanding that Thrupp & Maberly would be awarded the body order. A revised customer order for Chassis GXO71 was made out by Rolls-Royce on 17 September 1929 at a cost of £1185. This cost included 'extras' in the form of heavy duty 'Colonial' spoked wheels and polishing

*Golden Arrow. Body by Thrupp & Maberly. A 1929 land speed record holder*

the bonnet top. There were also two spare wheels, to be mounted on the front wings. An order for the body was placed with Thrupp & Maberly on 26 September 1929. Another late 'extra' was added by Guy. He had been told that the finest automobile clock was made by Jaeger in Switzerland and that they also offered a matching speedometer, manufactured to the same exacting standards. What he didn't know and probably didn't care about, was that the Jaeger speedometer was chronometric. Or clockwork as it was commonly called. This type of speedo demanded a speedo drive cable that rotated some 250% faster than the magnetic speedo, that the Derby factory normally supplied. So Rolls-Royce had to fit a small step up gearbox to drive the Jaeger speedometer. The sum total of the completed car with the extras came to £2,000. This was a huge sum - it would have bought no less than 20 Model 'T' Fords for instance. But to Guy it was money well spent, to get exactly the car he wanted. Three days after the order was placed, 'Black Friday' happened, when the bottom fell out of the American stock market. But it is doubtful if the great depression that followed, affected Guy financially; it certainly never showed.

Guy was now getting quite excited about his new car that was on order and started to take an interest in all things to do with the automobile. He learnt that one of the greatest annual motoring shows took place in London, at Olympia on 28 October 1929. Happily, this coincided with a business meeting he had to attend in London. Guy once again took a mini five day cruise on The Queen Mary and travelled back to the UK. Together with Walter Chapman, they visited the Olympia motor show. Once there, they made a beeline for the Rolls-Royce stand and were puzzled to see a new model on display. It appeared to be very similar to the 20HP chassis that Walter had ordered. But this chassis had a bigger engine fitted and was

GXO71 'Bluebell' being delivered by the coach builder to her first owner in 1930.

10

called the 20/25. Guy was not pleased and felt that Rolls-Royce had short changed Walter. He made his feelings clear to the Rolls-Royce people on the stand. Either they supplied him with a new 20/25 chassis, or he would cancel his order. By this time, GXO71 with the 20HP engine fitted had already been sent to Thrupp & Maberly. The new chassis was despatched from Derby by LMS steam train direct to the Cricklewood sidings on 25 October 1929. A temporary driver's seat was fitted by the Rolls-Royce Lillie Hall London Depot. The naked chassis was then driven the very short distance from Lillie Hall in Cricklewood Lane, to the Thrupp & Maberly works, also in Cricklewood Lane. While Thrupp's got on with building the body, Rolls-Royce at Derby commenced a new engine build. By early February 1930, GXO71 was back with Rolls-Royce at Lillie Hall, where 20/25 engine Serial No: O6Q was fitted in place of the original 20HP engine. All of this delayed the final trimming and painting of the body. The completed car was eventually delivered to Snoxalls Garage on 25 April 1930. Fred Snoxall registered it on Mr Mitchell's behalf. Which is why the car has a Hertfordshire, rather than a London registration - UR 6470. Parked on the forecourt of Snoxalls Garage, the new Rolls-Royce with its polished aluminium gleaming and sparkling in the early Spring sunshine looked absolutely awesome. Fred Snoxall considered it to be the prettiest car he had ever seen. And fitted with the bigger 20/25 engine, the car was no sluggard. Thrupp & Maberly had constructed the body from the same high grade, thin gauge metal, they had used on the Golden Arrow record breaker. The whole car 'on-the-road' weighed just 3580 pounds. In early summer, GXO71 pointed her shiny nickel plated radiator North on the A1 and headed for the Scottish Highlands. Behind the wheel and now dressed in traditional chauffeur's uniform, was Fred Snoxall. The garage was left in the charge of his eldest son Roy, while Fred drove Guy Mitchell and Walter Chapman deep into the Scottish Highlands.

There were just eight Thrupp & Maberly tourers, constructed on the small chassis Rolls-Royce.

| | | |
|---|---|---|
| GDK80 | 1924 20HP | Broken up. |
| GCK59 | 1925 20HP | Broken up. |
| GOK65 | 1926 20HP | |
| GZK81 | 1926 20HP | |
| GRJ80 | 1927 20HP | In India. |
| GLN39 | 1929 20HP | In India. |
| GXO15 | 1929 20/25HP | Broken up |
| GXO71 | 1929 20/25HP | 'Bluebell'. |

It was a fine 4 week's motoring holiday, that would be repeated annually, up until the outbreak of the second world war. Apart from occasional maintenance drives, the annual August holiday was the only time the car was used for the first 10 years of its existence. Guy had arranged for the car to be garaged and maintained on a permanent basis at Snoxalls Garage. Where it was parked in its own dedicated single car garage and covered in a dust sheet. Thrupp & Maberly were to go on and build just one more body to the GXO71 design, chassis no GXO15. This car was exported to the US, where the original body was sadly scrapped and a shooting brake body substituted. This car has not been heard of since 1977 and is presumed to have been scrapped. GXO71 (known as 'Bluebell') is now the only example on the small RR chassis, of what Thrupp & Maberly called their 'Phaeton' body.

By 1938 it was becoming obvious that war with Germany was inevitable. Guy realised that his idyllic motoring holidays to Scotland had to come to an end. Reluctantly he put Bluebell up for sale, with 40,000 happy miles recorded. He never ventured outside America again, for the duration of the war. However, with war clouds looming, an expensive fuel thirsty car like a Rolls-Royce was not in great demand. It was not until March 1939 that Guy found a buyer; albeit at the distressed price of £500. The new owner was a Mr Frears. He was the Managing Director and owner of Frears & Blacks Ltd, Abbey Bakeries, Leicester, UK.

J N Frears was a 'gentleman' engineer in the Victorian tradition and he soon modified the car to suit his own particular requirements. He lowered the front seat and altered the rake of the steering column. The fitted trunk which was probably showing signs of wear, following the many Scottish holiday was removed. In its place, he built a permanent 'trunk'. This trunk was constructed in the same fashion as the bodywork of the car itself; ash frame work paneled in aluminium. The top of this permanent trunk was finished in rolled aluminium to match the design of the body tub. The join where it met the existing body, was concealed by the hood. It may have endowed the car with a hugely increased load carrying capacity, but it looked hideous and later in the car's life, the

*Relative sizes: The 'small chassis' 20/25 and the 'large chassis' 40/50. Both bodies are by Thrupp & Maberly and are to scale.*

third owner removed it entirely. Mr Frears also discarded the unsatisfactory vacuum operated headlamp dipping mechanism and instead, fitted a valance mounted 'Pass Light'. During the period that Mr Frears was altering Bluebell, the country had declared war on Germany. But nothing much was seen to be happening and it was known as the 'phony' war. But behind the scenes, a lot was happening. Winston Churchill had at last been appointed Prime Minister and it was thought that if anyone could lead Britain successfully in a war with Germany, it was him.

In 1940 Mr Churchill was a worried man. In the few years since the declaration of war a lot had been achieved. Companies such as Rolls-Royce had performed production miracles and their most successful engine ever, the 12 cylinder 'Merlin' was coming out of the factories in ever increasing numbers. This engine powered Spitfires, Hurricanes, Lancasters, Tanks and a lot more. But Mr Churchill knew there was an even bigger threat to Britain, than the might of the German military that was now turning its attention to the conquest of the British Isles.

Churchill considered that it was entirely possible, that Britain would be starved into submission. This was because the country just did not have the land or manpower to feed itself. The UK was dependent on US grown grain. The problem was, that the ever growing might of the German U boat fleet under Admiral Doenitz, was sinking these cargo ships in mid Atlantic. More ships were going to the bottom of the ocean than were making it to the safety of a UK port. In fact, the German submarine fleet, were responsible for sinking 70% of all Allied shipping lost during the war years. Churchill gave the

*It is believed the driver in this picture, is a part time Chauffeur that Fred Snoxall employed on the owner's behalf. Fred (in the passenger seat) was not always able to leave his garage business for a whole month, to drive the owner & his friend on their annual motoring holiday to Scotland.*

'bread problem' top priority. A bigger and better convoy system was put in place, protected by an ever growing number of Royal Navy warships. For a while, this was successfully countered by the enemy, who organised their submarines into hunting 'wolf packs'. Churchill realised, that we had to make more productive use of the grain that was getting through from the USA. So the UK Government, head hunted a non-military man, who in essence was to be responsible, for putting the nation's bread on the table. What better man than someone who had been in baking all his life and was the owner of one of the biggest bakery's in Britain? Mr Frears, of Frears & Blacks, Abbey Bakeries, Leicester, was asked to take on the job. He accepted and was appointed as a Senior Civil Servant 'Head of Bakeries; Ministry of Food'. But it wasn't just Mr Frears that was conscripted. So was his car! Bluebell became his official Government car and was to put a lot of miles under her wheels during the war years. The car would obviously need further modifications; a highly polished motor car was not exactly suitable for the black-out restrictions that were in place in wartime Britain. So most of the polished aluminium was painted dark blue. The Flying Lady bonnet mascot was replaced with a plain cap and a blackout mask was fitted to the operational dipped headlamp. These masks were designed to show a glimmer of light on the road, but to block any light shining upwards and being seen by an enemy bomber.

1940 was a sombre period. The blitz was at its height and most major cities had been bombed by the Luftwaffe. Bluebell was now working hard for her living, but apart from some minor parking scrapes, she came through the war unharmed. Mr Frears new job, took him all over Britain. This was chiefly to visit bakeries and advise on production techniques, to

*Picture taken circa 1935. Standing by the pump is Fred Snoxall, who owned the garage that supplied and maintained the new Rolls-Royce. His son David is sitting on the bumper. David, now in his 80s, kindly supplied this picture.*

maximise bread output. Because of the importance of his work, Mr Frears was given an official petrol ration. But finding a garage outside of London that had petrol, could still be a problem. So Mr Frears dispensed with one of the two spare wheels and in its place, constructed a carrier that could be locked and held a large fuel can. Bluebell was now at war and during the conflict she covered 55,000 miles. Probably with little more maintenance than engine oil changes. Consequently when the war ended in 1945, Mr Frears sent Bluebell back to the Rolls-Royce Crewe works for a 'complete overhaul'. She was sent to Crewe, because the Derby factory by that time had been completely taken over by the Aeronautical Division of Rolls-Royce. The total cost of the overhaul was £350. In 1945 that would have purchased a modest three bedroom London semi-detached house! It was obviously a thorough overhaul and probably funded by the tax payer, but Bluebell had no doubt earned it.

In 1946 with 93,000 miles recorded on the speedometer, Mr Frears sold Bluebell to his brother, who was a doctor. His brother changed the plain radiator cap for a 'Spirit of Ecstasy' mascot, allegedly made out of solid silver. The doctor also sent the six wire wheels back to Dunlop where they were rebuilt with new spokes and repainted. Christmas 1947 was exceptionally cold and bleak. Rationing was in place, luxury goods unheard of and toys were in short supply. Russell, the doctor's son was just five and looking forward to Christmas day and his presents. The doctor had been lucky. One of his patients had alerted him to a toy shop which had received a limited supply of Mecanno sets. The doctor had loved Mecanno when he was a child; perhaps it was this that had

*Picture taken in the period 1940-1945. Bluebell has now been painted and a black-out mask fitted to the operational headlamp. She was the official car of a senior Civil Servant and racked up a lot of mileage during the war years.*

15

sparked his interest in all things mechanical in later adult life? The various size Mecanno sets were in austere 'Utility' packaging. The plain cardboard boxes had just the stock number stamped on them. The good doctor bought a complete set, wrapped the various boxes up and placed them around the family christmas tree. When young Russell opened the unadorned boxes and tipped the contents out, he was clearly mystified. What on earth were all these metal strips with holes? Not to mention the cogs, wheels and the box of nuts and bolts. The doctor was dismayed to see the disappointment on the face of his young son… but he had an idea. He hurried out of the house and into the garage, where he pulled the tool tray out from beneath Bluebell's front seat. Emptying the tools and debris onto his workbench, he gave the tool tray a quick clean and returned with it under his arm. With a theatrical flourish he scooped up all the Mecanno pieces and put them into the box. Which he then handed to Russell. The boy still did not know what the metal pieces in the box were for, but he did feel that he now had a real present. The box was beautifully made and even to a young eye, had 'presence'.

As Russell grew up, he came to love making things from Mecanno. In fact he became so attached to it, that when he spent some weeks with his grandmother every year, the box of Mecanno always went with him. In 1970 his grandmother died and when the loft of her house was cleared, prior to selling the house, the box was discovered. It was now tired and distressed; the Mecanno pieces rusted with flaking paint. The box was returned to Russell. With an adult's eye, Russell was intrigued to see that the box was not square, but trapezoid in shape. Further, he could just about make out a number stamped into the wood: 5235. He did wonder whether the box had

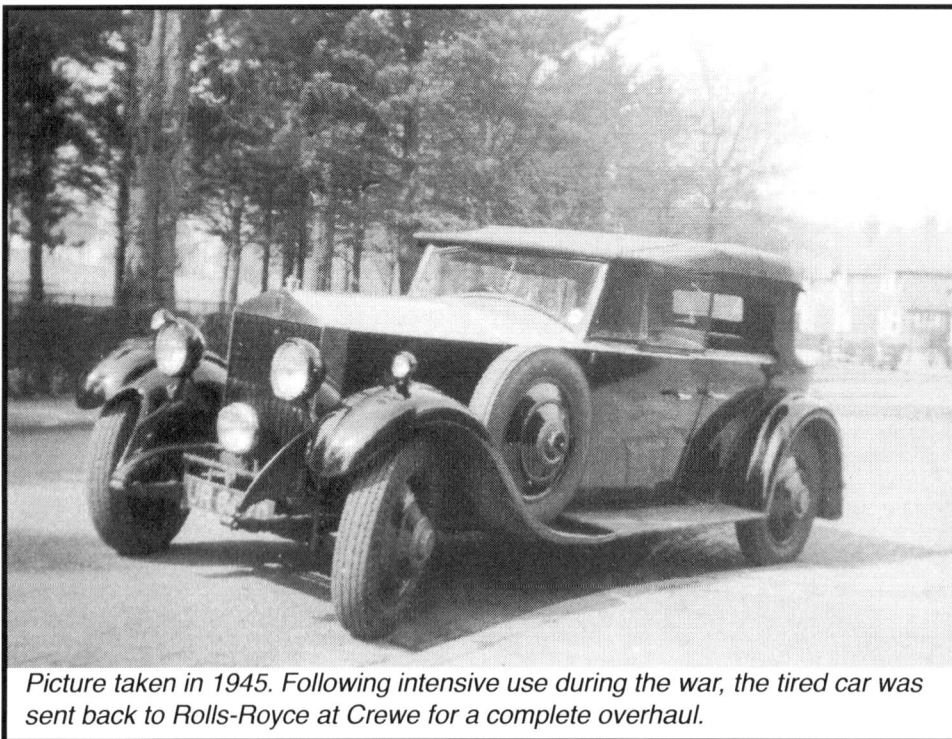

Picture taken in 1945. Following intensive use during the war, the tired car was sent back to Rolls-Royce at Crewe for a complete overhaul.

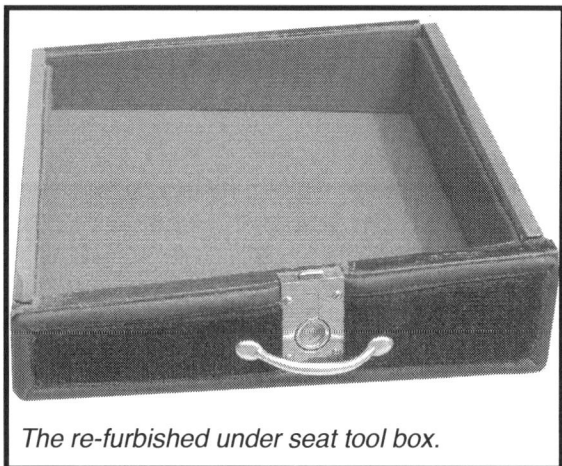
*The re-furbished under seat tool box.*

was the Thrupp & Maberly body number. Russell Frears very kindly returned the box to me and which is why, 61 years after it had been removed, the original toolbox was reunited with the car. Alas, the tools that it originally contained were never seen again.

Doctor Frears used the car until early in 1954. It may well be that maintenance, while in the doctor's ownership, was not all that it should have been. Because the engine began to make itself heard and started to smoke. The doctor stopped using the car and it sat neglected in his garage. It was not until the Spring of 1960 that the doctor decided to get rid of the car and Bluebell was put up for sale. This decision proved to be very fortunate for the car, because the next owner was a Rolls-Royce master

anything to do with his father's old car. But the number meant nothing to him and neither did it match the car's registration, clearly visible in an old photograph he had. The faded box with the rusting Mecanno was put into his own loft as a keepsake and over time was once more forgotten. Some 35 years later, the current owner of Bluebell tracked Russell down and asked if he had any photographs of his father's old car. On impulse, Russell asked the new owner if the number 5235 meant anything to him. It did! It

*Picture taken in the 1950s. Mr Frears the owner (then) has added a rear 'Auster' screen. These screens give a surprising amount of wind protection to passengers in the rear seat.*

mechanic. The low price the new owner paid for the car, reflected its poor mechanical condition. The doctor's final act, was to remove the silver Spirit of Ecstasy mascot, with the intention of selling this separately. The plain cap the car had worn during the war was pressed back into service and Bluebell was sold. The buyer was Ken Hucknall, the owner of a small garage in Nottingham.

Ken Hucknall was a professionally trained engineer. When war started he was conscripted into the RAF where his mechanical skills were put to good use on the engineering maintenance side. It was during the war that Ken developed a lifelong passion for all things Rolls-Royce. The more Merlin engines he stripped and overhauled, the more he appreciated the quality of the design and the engineering excellence they displayed. By the time the war came to an end, Ken had risen to the rank of Wing Commander and was the Commanding Officer of a Repair & Maintenance facility. With the end of the war, Ken found himself demobilised with a gratuity. Using this seed capital, Ken acquired a small run down garage in Forest Road, Nottingham called 'Ristes'. This garage was little more than a general car workshop and no different to hundred's of similar country garages, that had grown out of early bicycle workshops. But Ken was to change Ristes out of all recognition. Using his Rolls-Royce wartime expertise, Ristes started

*Picture taken at the end of World War II in Western Park Road, Leicester, the home of the third owner Doctor Frears. The picture shows the rather ugly rear mounted boot extension that the second owner (his brother) built himself, to match the lines of the car. Also visible on the offside wing is an external carrier. It is thought this was to carry a spare fuel can, during the war years.*

to specialise in the maintenance and care of Rolls-Royce and Bentley motorcars. The business prospered and today, Ristes are one of the leading Rolls-Royce car restoration workshops in the UK.

When Ken saw the advertisement for GXO71 that had been placed by Doctor Frears, he went to look at the car. He bought Bluebell on the spot and had her transported to Ristes. This was on 22 March 1960. Bluebell by this time had 150,000 miles recorded on the speedometer, plus another (estimated) 2,000 miles had been run with a broken speedometer cable. By the following September, a major overhaul had been completed. Ristes had re-bored the engine and fitted new pistons, gudgeon pins and small end bushes. Ken Hucknall recorded that this was the first rebore in the life of the car. Ristes also fitted new valve guides, springs and new bearings in the timing gears. The crankshaft was unworn and the original big

end bearings were fettled and refitted. The brakes, clutch and servo linings were all replaced. The prop shaft was reconditioned and new front wheel bearings fitted. The water pump, dynamo & starter were all overhauled. A new double duck black hood and hood bag were also fitted. Apart from putting the car back into good order, Ken also moved the steering column back to the original position and discarded the front T B André bumper. This had become bent and distorted from parking mishaps and Ken decided the car looked better without it. Bluebell still had her

Ken Hucknall, who died in September 1972.

Picture taken in 1960 outside a pub in Nottingham, when Bluebell was Ken Hucknall's personal car. Ken had acquired 'Ristes' after the war and transformed the business into a leading Rolls-Royce & Bentley restoration centre. Steve Lovatt the current owner of the business, was Ken's first apprentice.

Picture taken in 1960. Engine overhaul underway at Ristes.

wartime blue paint finish; this had been refreshed in late 1948 when a new coat of clear varnish had been applied. The paintwork still looked good and as Ken didn't know that polished aluminium was the original finish, Bluebells's polished magnificence was to remain hidden for very many more years. Ken also sourced a correct Spirit of Ecstasy mascot to replace the plain cap on the radiator. Bluebell was not only once more looking splendid, she was also back in first class mechanical order. It is clear that Ken intended to keep Bluebell for his own use, as he had transferred his own cherished registration number KH 21 onto the car. But possibly his fledgeling garage business needed some cash and Ken decided to take the profit represented by the Rolls-Royce. Ken contacted Commander Hugh Keller of Paddon Brothers, who specialised in the marque.

Paddon Brothers of 60 Cheval Place, London, SW7 were commissioned by Ristes to sell the car in the autumn of 1961. Commander Hugh Keller who was the Sales Director at Paddon Brothers had world-wide sales contacts. It did not take Hugh long, in finding an enthusiast in New York, who was willing to pay the asking price. As part of the sale negotiations, Paddon's converted the headlamps to dipping double filament bulbs and also converted the tail lamps to double filament to operate as stop lights. The wheel discs were discarded, as it was considered that the car looked 'sportier' without them. Paddon's also removed the bulb horn as this was now considered old fashioned; in any case, the car did have an electric klaxon horn.

Bluebell was acquired in October 1961 by Alec Choremi of Locust Valley, New York State, USA. He paid £750 for the car, exclusive of shipping charges. However, the car was not delivered to the USA for another six months, because of the amount of additional work that the new owner requested. It was also the case that the winter of 1962 in New York was particularly harsh and Alec Choremi did not plan on using the car until the spring. Mr Choremi visited Paddon Brothers in London

Picture taken in 1960. Engine overhaul underway at Ristes.

and viewed the car in January 1962.

Following that visit, Paddon's reupholstered the front squab and supplied new front and rear tonneau covers. They also refurbished the running boards and billed Mr Choremi for welding new aluminium repair panels into the rear of both front wings. However no evidence of this repair was found when the wings were removed in 2004. The lower half of the car was repainted, a new offside glass fitted to the quarter light and the headlamp rims and front screen (incorrectly) chromium plated. At Mr Choremi's request they also fitted a reconditioned RR thin rimmed steering wheel. The original steering wheel was later sold in the USA. All six wheels were sand blasted and stove enameled. Two new tyres, clutch and brake pedal rubbers plus a spare ignition key brought the 'extras' bill up to £290. The car as delivered and including shipping charges, cost Mr Choremi a total of £1150. This was a considerable sum in 1962. To put it in perspective; circa £3500 would have bought a brand new detached

three bedroom house in Sussex. Bluebell was eventually despatched from Southampton Docks on 26 April 1962. The car travelled to America in the Steamship SS Sidonia and was delivered to the General Standard and Marine Corporation, 21 West Street, New York City.

Soon after Alec Choremi received the car, he was contacted by Doctor Frears to see if he wanted to buy the solid silver 'Spirit of Ecstasy' that he had removed as a keepsake. He wanted £50 for the mascot. The offer was refused. As the car possibly looked a little dull to American eyes, with its dark blue paint finish, Alec Choremi soon added a yellow body stripe. He also removed the original headlamps and fitted bigger Lucas P80 headlamps, together with matching front side lamps; all finished in

*Picture taken in 1961 in New York State. Mr Choremi the new owner has added a yellow body stripe to the car. Note the chassis number used as a number plate.*

original nickel silver and sourced from Paddon Brothers in the UK. Again, at some expense! New running board metal edges and new wing piping was also ordered from Jack Barclay Ltd in the UK. After spending all this money on his new toy, Mr Choremi ran into a little business difficulty and he had to raise cash. Reluctantly he advertised Bluebell for sale in 'The Flying Lady', which is the US Rolls-Royce club's monthly magazine. He found a buyer who already had a huge Rolls-Royce and Bentley car collection.

Almost as soon as the car was sold, Alec Choremi regretted it. It was not more than a couple of months when his business fortunes changed again and he wrote to Charles Mallory the new owner. Mr Choremi asked for first refusal should Mr Mallory ever want to dispose of the car. When indeed this happened, Mr Choremi's business was once again short of cash and he was not in a position to repurchase the car and it was sold to Carl Malmaeus. Mr Choremi later pursued Mr Malmaeus in an attempt to reacquire, but terms could not be agreed.

Charles Mallory of Stamford, Connecticut, acquired the car on 1 September 1977 and paid Mr Choremi $20,000. Mr Mallory is an avid automobile collector who has owned 'hundred's' of classic & vintage cars. In a letter to the current owner he stated 'the car holds a special place in my heart'. It would appear that shortly after acquisition, Mr Mallory drove the car too hard and fast and destroyed a big-end bearing. Consequently, the engine was completely rebuilt in 1978 by Herb Lubarsky of 21 Swol Street, Chicopee, Ma 01013 at a cost of $5,300. Lubarsky was a noted American specialist engine restorer. During this rebuild, the engine was converted to the later specification of

Picture taken in 1983 in Connecticut, USA. Mr Mallory the new owner has repainted the car two tone blue, with a polished aluminium bonnet.

a 5.25:1 compression ratio and which gives a small increase in power. In 1978, Charles Mallory also had the seats re-upholstered in black leather to the original design and ordered a wings off complete repaint, using cellulose. Mr Mallory also replaced the black (and presumably worn) tonneau covers and the hood, with stone coloured double duck canvas, which was the original colour. In spite of spending large sums of money on the car, Bluebell was not used much during Mr Mallory's ownership. Because he had so very many cars, he couldn't drive them all and most became essentially museum pieces.

Carl Malmaeus acquired Bluebell in June 1986. At this time the vintage car market in America was depressed and he negotiated the acquisition for the bargain price of $14,000. Carl had actually visited Mallory's car collection to view a vintage Bentley 4.5 that was for sale. However, the Bentley did not meet Carl's expectations and on a whim he test drove Bluebell. He later stated that the Bentley

lived up to its reputation as a 'fast lorry' and that when he then drove Bluebell, the Rolls-Royce was so silky smooth in comparison, he decided on impulse to buy her. Part of the reason, was that he suspected that the price would never be this low again. Carl by the way, is a Banker and perhaps unsurprisingly, is financially astute! Bluebell was delivered from Connecticut to Hudson New York, where Carl lived at that time. Later his Bank sent him to open and develop a new branch in Toronto, Canada. By this time Carl had sufficient confidence in the car, that he drove her the 500 miles to his new home. He remarked afterwards that there had been no mechanical problems with the car and that it was one of the nicest drives in a vintage car he'd ever

*Picture taken in 2004 in Sweden, when the car was owned by Carl Malmaeus. Behind the car is the farmer's barn where Bluebell was stored.*

experienced. In 1987 Carl was moved once again by the Bank and asked this time to repeat the Toronto success in Croatia. This recently war ravaged country was no place, in Carl's view, for a valuable vintage Rolls-Royce. Consequently, Bluebell was shipped back to Torekov in Southern Sweden, where Carl has a summer house and garaged in a nearby farmer's barn. Bluebell was not driven a great deal during Carl's 18 year ownership. A 'J type' Laycock overdrive unit was ordered from the UK by Carl in about 1992 and which cost £600. A local Swedish garage fitted the unit, but Carl considered the work unsatisfactory. The work was redone twice before the overdrive worked to his satisfaction. The overdrive is a boon and had they been available in 1929, Henry Royce would no doubt have fitted them as standard.

Carl also did lots of cosmetic work on Bluebell. The engine bay was properly cleaned and all the appropriate parts that were originally stove enameled by Rolls-Royce were removed, blasted and re-enameled. Most of the pipe work within the engine bay was also replaced and the starter motor and dynamo, removed and re-nickeled. The chassis was steam cleaned and repainted using chassis black. The leather trim on the inside of the doors was replaced and the door hinges were re-polished. Much of the rest of the bright work was also stripped of its incorrect chromium plating and re-plated in German nickel silver, including the radiator shell. The side quarter light window frames were remade in yellow stainless steel to match the rest of the nickel bright work. A new Rolls-Royce coil and rotor arm were fitted in December 1993; also the

*Picture taken April 2004. Still displaying her Swedish number plate, Bluebell waits in the ferry queue at Gothenburg Docks. After 43 years abroad, she was finally coming back home to the UK.*

central chassis lubrication system was overhauled and some pipe runs replaced. Flashing indicators are a legal requirement in Sweden on all cars. Consequently, Carl had the original side and rear lights converted and had a flasher switch discreetly fitted at the bottom of the dash. Some minor aluminium welding work was carried out on the under trays; all of which are original and still in place. Carl mentions in an e-mail that oil is still not getting to the rear shackles on the rear springs. This is actually a design fault and the solution is to fit separate oil nipples on the rear shackles. This has now been done. Finally, Carl removed the paint from the top of the scuttle and door tops as he considered the car looked better that way. He wasn't to know, that was the way the car had originally left Thrupp & Maberly in 1930. In 2004 Carl, with retirement on the horizon, decided to sell Bluebell and replace her with a vintage WO Bentley. He placed a small lineage advertisement in the British Rolls-Royce Enthusiasts Club

magazine. The car was described as being in Sweden, consequently there was limited response. The current owner emailed Carl with the first of what was to eventually become nearly 50 messages and replies. Not wanting to travel to Sweden to inspect what might have been a rubbish motor car, I asked Carl via email detailed questions. At the end of this barrage, Carl decided that the enquirer was not a time waster while I had formed the opinion that Carl was a man of integrity. I flew to Copenhagen with my wife Joan and we travelled onto Sweden by train. It was love at first sight when the doors to the barn were opened and the car was revealed. I was immediately taken with the striking lines of the car, little knowing then just how rare the car was. Nor did Carl

*David Snoxall (in the car) with the author in June 2008. David is the little boy seen sitting on the front bumper in the picture on Page 14.*

have any knowledge then, of the car's provenance.

On 7 April 2004 I purchased the car in Sweden. Carl arranged with his local garage, that they would allow me to use their facilities. I needed to give the car a thorough check, as she hadn't been used much for a long time. I changed the engine oil, topped up all other oil levels, checked the tyre pressures and headed for the Swedish port of Gothenburg. En route the dynamo ceased to charge and there was a partial fuel blockage that took a few miles to clear itself. But the journey was largely uneventful (but so cold!) and the drive to Gothenburg was all joy. We overnighted in a large hotel which had secure parking and I bought a spare fully charged battery to get us home, after discovering that the magneto wasn't working either. We sailed the following day to Newcastle and just managed to get back home to Sussex, before we needed lights. Although cold and tired, the first thing I did was to change the engine oil again, while the engine was good and hot. I had been forced in Sweden to put a modern multi-grade oil in the engine. This was drained and the sump filled with Castrol Classic 20/50.

I needed to find out a lot more about this vintage car. The first step was to order a copy of the chassis build cards from the RREC. Immediately I was confronted with a puzzle. The first owner's address was in the West End of London, but the car had been delivered and registered in Boxmoor, Hertfordshire.

*Bluebell on tour in France, at Mont St Michel.*

So I wrote an article stating I was trying to trace the roots of this old car and sent it to 'The Hemel Hempstead Gazette'. The Gazette is the local newspaper in the Hertfordshire catchment area where Snoxalls Garage, the supplying retailer had been located. I got a reply from David Snoxall from whom I learnt the fascinating early history of the car. I also heard from Tom Clarke, the Rolls-Royce historian and author. Tom had been made aware of an old archive of automobile photographs that was held by 'International Automobile Photos' of Box 1028, Minneapolis, USA. He acquired access to part of this archive and discovered many images produced by Chas Bowers, a British professional car photographer.

Bowers was responsible for many of the Thrupp & Maberly pictures taken in the late 1920s. Tom also had access to Bowers' original ledger, detailing when and where the pictures were taken, together with chassis and body numbers of the cars. In this treasure trove of glass negatives, was an image of Bluebell when new, taken on 10 April 1930 at The White Lodge, The Avenue, Hampstead, London. This was probably taken when Thrupp &

Maberly was delivering the new car.

For the first time I could see what a magnificent car this had been. And I also now knew what had to be done visually, to restore the car back to its original condition. In 2005 I dabbed paint stripper onto a small part of a rear wing. With three applications the paint removed cleanly with no scraping required. Now revealed for the first time in over 60 years, was polished aluminium. Wow.

It felt like I'd struck gold!

# Part Two

# Mechanical & Maintenance

*All pre-war Rolls-Royce cars are different. Even chassis that rolled out of Derby in the same week, will have at the very least, dimensional differences, as they were all hand crafted. Rolls-Royce were also continually developing and improving their product all the time; this process never ceased. On some cars for instance, it is possible to dismantle the fuel tap, without removing the Autovac. On other cars, there is not enough clearance to do this. So you must bear in mind that much of the technical information that follows, relates to GXO71 and you may well have to adapt it to suit your own particular chassis. The following information and guidance is not intended for the experienced Rolls-Royce mechanic, but more for a private owner of the marque, who is not afraid to get his hands dirty and use a spanner. So read on for the detail, of the problems encountered and the solutions found, and which finally resulted in a totally reliable car. It is hoped my experiences will assist in some small way, other owners. Where appropriate, I have also included some information from dealing with mechanical issues on my 25/30. This chassis was a development of the earlier model and has a close similarity.*

The 'tappets' refers to the clearance that needs to exist, between the push rod and the rockers that operates the valves, in the cylinder head. When out of adjustment the tappets make a distinctive 'tapping' sound; hard to describe, but very recognisable.

Clearances for all valves is .004" measured with the engine cold. To adjust, first remove all six spark plugs, to make the engine easier to rotate. Slacken right off, the three knurled nuts on top of the rocker cover and remove the cover. On some cars, this will also mean that the air cleaner/silencer that is positioned on top of the engine, will also have to be removed. It is essential when adjusting, that the valve you are working on, is completely shut. This is easy to get wrong, because the cam that rotates and pushes the push rod up to open the valve has 'quietening ramps'. Which means that for almost half of its rotation, it is lifting (or closing) the valve slightly. The simple way of ensuring that the valve is completely shut, is to adjust each valve, following the table on this page. If you are working single handed, a useful tip is to secure a pencil or similar, with a bulldog clip, to the rocker operating the valve. Turning the engine over slowly on the starting handle, you will be able to easily see the pencil dip, when the valve closes.

On earlier models, the tappet adjustment is on the *bottom* of the pushrod and accessed by removing the thumb screws on the tappet chest covers. Note that the locking nut is the bottom nut in the tappet chest and that the top (adjuster) nut should be screwed up to tighten the valve clearance and down to increase the clearance. The lock nut is screwed down - ie: turned clockwise to tighten. On later models, the tappet adjustment is at the top of the engine, on the rockers operating the valves. Which makes adjustment a lot easier.

Following reports from the 20HP section of cylinder head nuts coming loose and blowing the head gasket, it is advised that these nuts are checked for tightness (torque to 18/20 lb/ft) when doing work on the tappets.

## Adjusting tappets

| Adjust this valve | When this valve is open |
|---|---|
| (No. 1 is nearest to the radiator) | |
| 1 | 12 |
| 2 | 11 |
| 3 | 10 |
| 4 | 9 |
| 5 | 8 |
| 6 | 7 |
| 7 | 6 |
| 8 | 5 |
| 9 | 4 |
| 10 | 3 |
| 11 | 2 |
| 12 | 1 |

When faced with an engine that is not running as it should, bear in mind that carburation and ignition are closely related. It is often the case, that a perceived carburation problem, is actually an ignition problem. So before you turn your attention to the carburettor, first make sure that every component in the ignition system, is performing correctly. Automatic control of the gas and the spark on the early 20/25 Rolls-Royce is rudimentary, in comparison with a modern car. Hence the mixture, hand throttle and early/late controls on the steering wheel, need to be understood and used, by the driver.

The hand throttle should be set by ear, to give a reliable steady tick-over. As the engine comes up to temperature it gets more efficient. Hence the revs will rise and the hand throttle needs to be reduced a notch or two, back to a steady tick-over. To achieve a

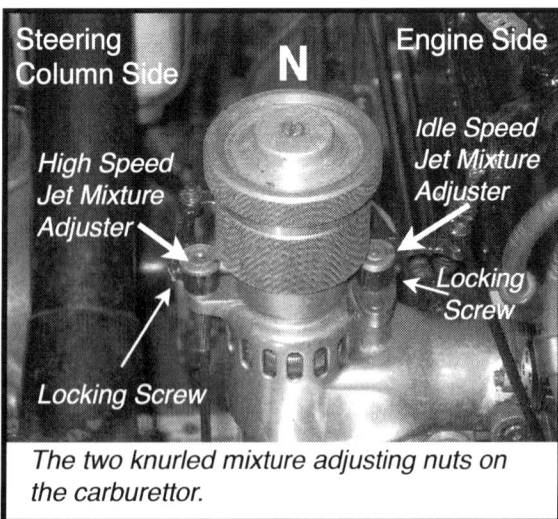

The two knurled mixture adjusting nuts on the carburettor.

really slow idle, when hot, the ignition early/late control also needs to be moved, from its running fully advanced position (control fully up), to half retard (control in mid-way position). You will also find, that the mixture control needs to be moved (to the right) 2 or 3 notches rich. All of the above, pre-supposes that the carburettor has been correctly set up. The detail on how to do this, is spelt out later in this section.

Before embarking on adjusting the carburettor, it is useful to have a basic understanding of how the twin jet, Rolls-Royce designed, carburettor operates. Like all carburettors, the supporting fundamental, is the Bernoulli principle. This states, that as the velocity of a fluid increases, the pressure exerted by that fluid decreases. In the case of the carburettor, the 'fluid' is air. As the pistons in the engine descend on their intake stroke, they create a vacuum, which sucks air through the carburettor. Inside the carburettor, is a narrowing of the throat, called a Venturi. This constricts the air flow and therefore speeds it up. The fast moving air is channelled under the air valve, where it's varying pressure drop, depending on throttle position, is used to move the air valve against the action of a spring. The air valve is connected to a disk, that controls the size of the throat, that the high speed jet operates in. The correct poundage of the spring is critical, to this operation. The low speed jet is not controlled by the air valve.

Consequently, it is more important to get the mixture adjustment absolutely right on this jet, as it is in operation at all speeds - not just when the engine is idling. The jet needles have a conical control section, the needle & cone operating inside the jet tube. Petrol flows within these tubes. As the needle is moved up, it allows more petrol to exit at the cone. The petrol emerges as a spray, straight into the fast moving air flowing through the carburettor. This air flow atomises the petrol into a gas; the quantity of gas allowed to pass into the engine to be burnt, is controlled by a butterfly valve, operated by the throttle pedal.

All 20HP and early 20/25 cars have the Rolls-Royce designed twin jet carburettor. There are various series of this carburettor, but operation is essentially the same for all of them. Later cars were fitted with a simpler SU type carburettor, which is outside the remit of this book.

### Annual maintenance

The key issue in servicing this carburettor, can be summed up in one word - cleanliness. Especially the air valve & piston, which should be removed and cleaned in petrol using a lint free rag. Remove the air intake and clean out any carbon deposits within. Do not be tempted to polish *any* of the carburettor internals. Ensure that the captive spring (see picture 'C') underneath the high speed needle is free, as this maintains constant upwards pressure on the needle. The jet control arm

bears down on the top of the flat conical tip at the end of the needle, against the tension of the spring and is finely controlled by the knurled mixture adjusting nut. If you have had to replace the jets, or had the carburettor in pieces for whatever reason, then you need to set the initial adjustment on the jets. This is done, by screwing the knurled adjuster down, until the top of the control rod is flush with the top of the knurled adjusting nut. The knurled adjusting nuts should then be adjusted, with the engine running, as outlined later in this section.

If you accidentally knock the starting carburettor control to 'ON' while driving, the engine will run rough and probably die on you. On my car, this control could also shake itself on, when driving on very rough roads. To ensure this does not happen, it is

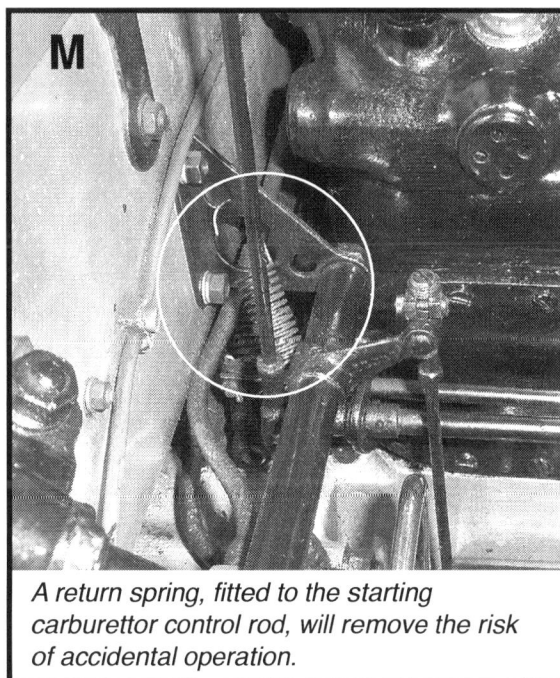

*A return spring, fitted to the starting carburettor control rod, will remove the risk of accidental operation.*

recommended that a weak return spring is fitted, as shown in picture 'M'. This will make sure the control is always either positively on, or positively off. Later cars were fitted from new with a return spring.

If the car has not been used in a very long time, then drain the petrol from the tank and the Autovac and discard it. Modern fuels have a relatively short shelf life and after six months or so, begin to lose their volatility. Did you know, that aviation fuels *have* to be discarded after three months? This 'old' kerosene is actually mixed with fresh and then enters the domestic oil market. I found this out, when our Aga started playing up. The Service Engineer who called, turned out to be an ex British Airways engineer, who had taken redundancy.

### Starting carburettor

This is a small self contained carburettor built into the main carburettor and can be thought of as a choke. Use it for starting, when the engine is completely

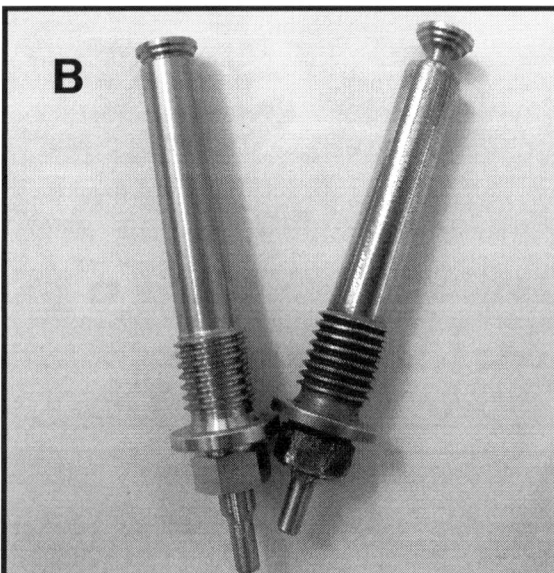

Starting carburettor needle and jet removed for cleaning.

The new high speed needle & jet (Left). The needle slips in freely to this position. Compare on the old unit (Right) the position of the needle. At certain rotational positions, it had to be pushed firmly to get it this far down in the tube. In service, the spring tensioning the needle (picture 'C') would have had a job to control the 'sticky' needle. Hence the erratic running of the engine.

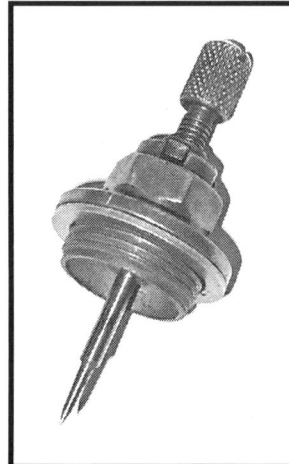

Starting carburettor (arrowed).

cold; on a hot summer's day it should not be necessary to use it, for more than a second or so. To start a cold engine, completely close the steering wheel idling control and hold the starting carb control fully on. As soon as the engine starts, back off the starting carburettor, while at the same time bringing the idling control about two thirds of the way up, to a fast idle. If the engine hesitates, then open the starting control a little more, until the engine is happy to run with the starting control fully off. Once a year remove the jet/needle and clean, but do not oil. Screw the jet down until you feel it bottom and then open by one & a half to one & three quarter turns, depending on whether it is summer or winter.

If the engine hasn't been started for a week or more, it is beneficial to unscrew the float chamber top by four turns or so, until petrol spurts out of the overflow. Then tighten back up. This procedure will avoid the engine turning over on the starter for a long period.

### Tuning the carburettor

First take the car for a long run and get the engine hot. Does it need adjusting? If you find the slow idle is better when the mixture control is moved away from the centre position, *in either direction*, then unless this is a small one or two click movement of the control, the mixture does need adjusting. Set the early/late control in the middle of its range. Slow the engine on the hand throttle, to the point that one more click and the engine will stop. The aim is

for a very slow tick-over, without the air valve lifting off its seat. Leave the engine ticking over and lift the offside bonnet. Refer to picture 'N' in this section - the slow speed jet is the one nearest the engine block. Slacken the pinch screw that locks the knurled nut. Weaken the slow running mixture, by turning the knurled nut slowly, anti clockwise. You will not turn it far, before the engine runs rough, or stalls. Mark this position on the knurled nut with a felt pen. With the engine again running, repeat the exercise, by strengthening the mixture; ie turn the knurled nut the other way. Once again the engine will run rough, or stall. Mark this position. Then turn the knurled nut to exactly mid-way, between these two extreme positions. Tighten the pinch screw to lock the adjustment. Get behind the wheel again and with the engine ticking over, move the steering wheel mixture control, slowly all the way to fully weak

*Domed nut with secured spring (inside) situated at the bottom of the carburettor. This tensions the high speed needle. The spring, inside the nut at at **C**, should be clean and free with 4mm of up & down movement.*

and then to fully rich. If the mixture setting at the carburettor is correct, then moving the mixture control as indicated, will result in the engine begin to run rough, or even stop, at the extreme ends of the mixture control range.

*The procedure to set the high speed jet is exactly the same as you did with the slow speed jet; finding the mid point of adjustment on the knurled nut.* The high speed jet is the one nearest to you, as you look at the carburettor. However, the high speed jet does not function, until the air valve lifts. This equates to about 4mph in first gear. So set the steering wheel early/

late control and the hand throttle, fully up. The engine will now be running quite fast and it is at this speed, that the high speed jet is adjusted.

Take the car for a road test, with the mixture control set in the middle position and the advance & retard (early & late as RR prefer to call it) set fully advanced - control fully up. At a steady 30-40mph in top gear, move the mixture control slowly fully left and then slowly fully right and observe the effect on the car's performance. If the best running is achieved with the mixture control either in the middle of its travel, or at best no more than 2 clicks either side of centre, then the carburettor is correctly set up. If you do need to make an adjustment on the road, do it in small steps. Picture the knurled adjusting screws as clock faces. Each adjustment made, should represent no more than 2-3 hours of movement.

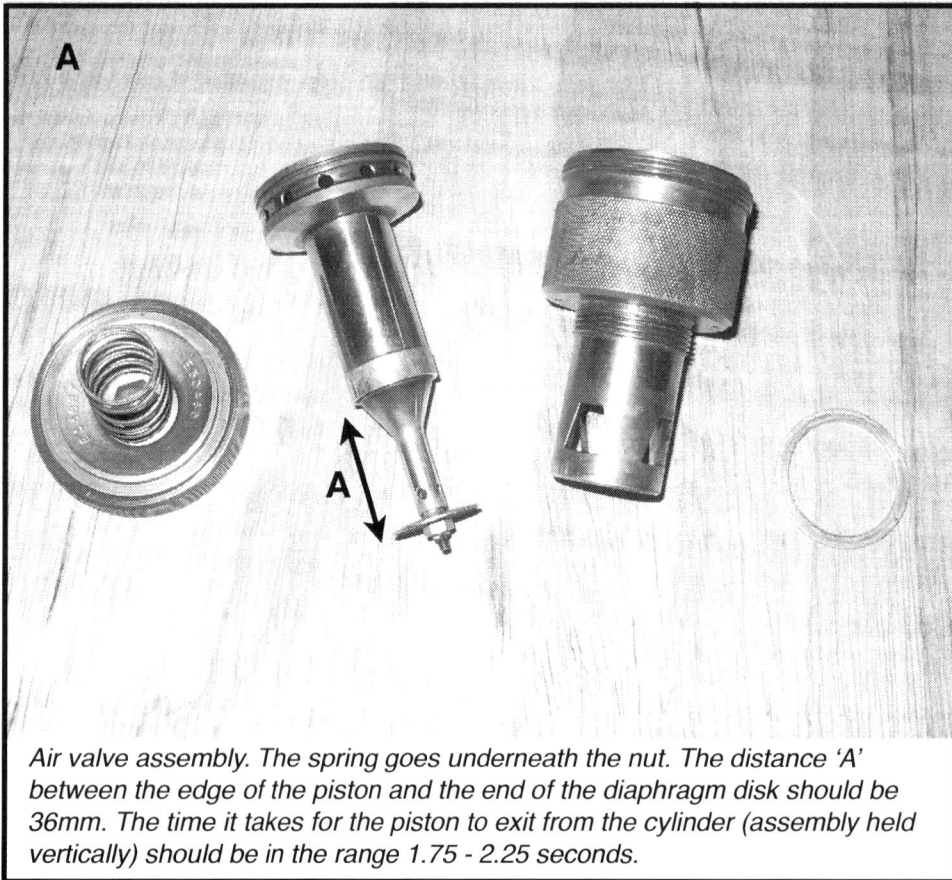

*Air valve assembly. The spring goes underneath the nut. The distance 'A' between the edge of the piston and the end of the diaphragm disk should be 36mm. The time it takes for the piston to exit from the cylinder (assembly held vertically) should be in the range 1.75 - 2.25 seconds.*

You will discover that the weather, the altitude and the ambient temperature will all have an effect on the way the engine performs. With a correctly set up carburettor, altering the mixture control a few notches either way, will compensate for these changing factors, as you drive. Use your ears and 'feel' what the engine is doing!

Do not expect an early car to always run perfectly. Going downhill on a trailing throttle, some popping on the over-run can happen, even with a perfectly adjusted carburettor. This is because when you lift your foot off the throttle while the air valve in the carburettor is in use; in other words lifted, the air valve immediately drops, resulting in an over-rich mixture very quickly. This is then dumped into the exhaust manifold where it ignites and 'pops'. If you were able to observe the air/fuel ratio by means of an external

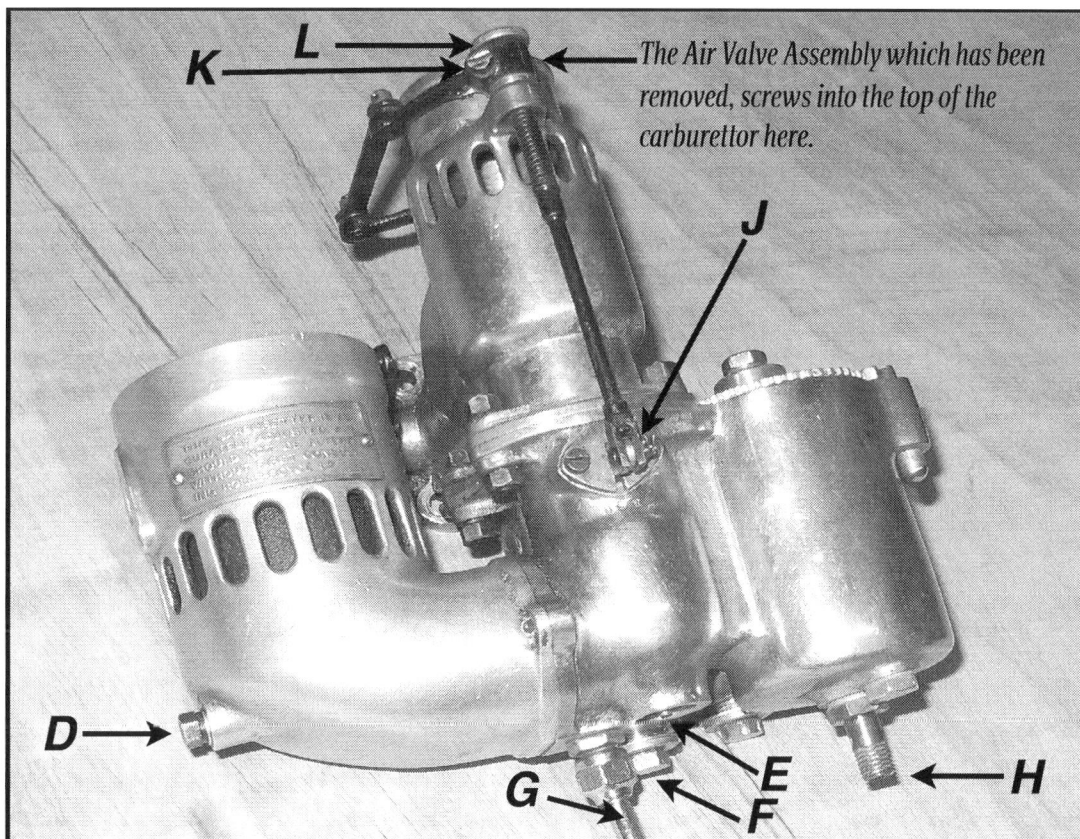

*The Air Valve Assembly which has been removed, screws into the top of the carburettor here.*

*L: Knurled screw to adjust high speed mixture. **K**: Locking screw. There is a similar arrangement on the other side of the carb, to adjust the low speed mixture. **J**: Remove split pin and clevis pin and undo two screws either side. Jet actuating lever can now be removed and needle removed from top of the carb after unscrewing air valve assembly **D**: Single bolt to remove air intake. **E**: Nut with tensioning spring has been removed, to show the tip of the high speed needle. A socket spanner here, allows the needle and jet to be removed downwards while the carb is still bolted to the engine. **F**: Access to the low speed jet & needle. **G**: Overflow drain tube that exits just above the engine under shield. **H**: Fuel inlet pipe connection.*

instrument, you would see it drop from it's stoichemtric ideal of about 14.7 to around 10 (very rich). If you do nothing, it will slowly increase again to 14.7 or thereabouts. This is because movement of the air valve does not happen quickly enough, to mirror the changes in air flow. The mixture similarly becomes too weak, when you suddenly accelerate and it takes a moment to catch up. You can mitigate the former effect, by retarding the ignition and weakening the mixture as you lift off the throttle, going downhill.

From the above, you will appreciate that the controlled movement of the air valve is very important. The spring that does the controlling needs to be in good shape. You can check this, by unscrewing the air valve from the carburettor. After removal, turn the valve assembly upside down over your hand and time how long it takes, for the piston to fall. It should be in the range 1.75 - 2.25 seconds. If it falls freely, then check that the top cover is making a good seal with the top face of the cylinder. Lap the cylinder to the top face with Brasso or similar. With the piston in place in the carburettor body, the spring that is fixed to the top cover, should lift the cover by about 1mm, before being compressed when the knurled holding ring is screwed down. A different result to this and you should fit a new spring; available from Fiennes at under £20. Note the series of holes at the top of the air piston in picture 'A'. Your piston may have a different

number of holes; perhaps only 1 or 2. The piston could be made from aluminium or brass - both were used by the factory at different times. These holes are used to fine tune the speed of the piston when in use and which is why, it is not a good idea to swap pistons with another carburettor.

I had a bizarre carburettor problem on Bluebell. I would carefully adjust both jets and the engine would run well. But after just a few miles, the mixture would be all over the place and the car would pop and

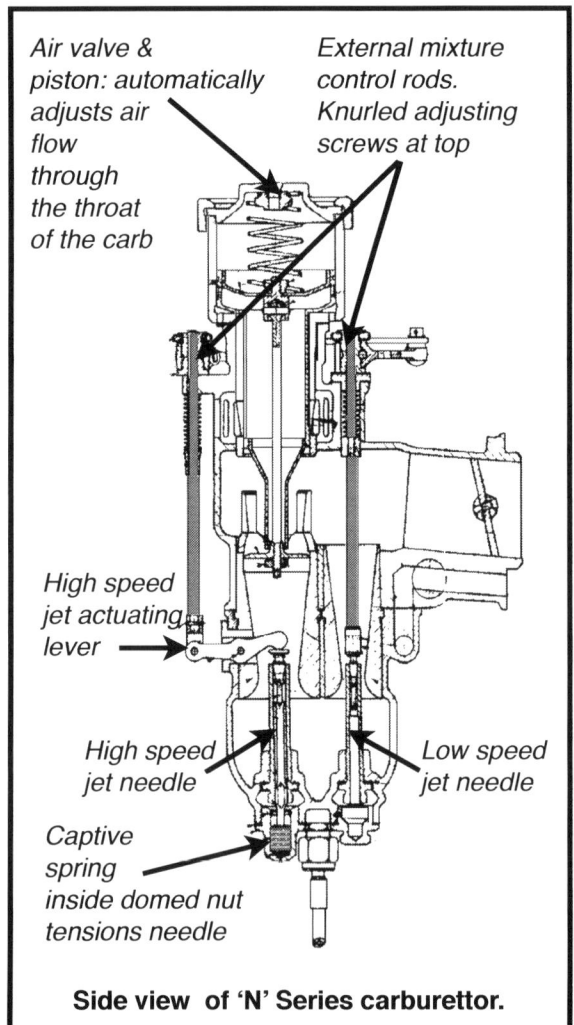

Air valve & piston: automatically adjusts air flow through the throat of the carb

External mixture control rods. Knurled adjusting screws at top

High speed jet actuating lever

High speed jet needle

Low speed jet needle

Captive spring inside domed nut tensions needle

**Side view of 'N' Series carburettor.**

36

bang in the exhaust, on the slightest of downhills. On a major hill it would backfire so violently, I feared for the exhaust system. All was clean internally and the air spring had been renewed. I wrestled with this problem for weeks. Until one day I had a bright idea. I removed the high speed jet and needle complete and with them in front of me, rang Fiennes. I asked Rob in the Parts Department, to get the same (new) components in front of him. He then described the action of the needle going into the jet. The needle slipped freely into the jet until the flat section was flush with the top of the jet. My needle would do this - but only at certain rotational positions. At other positions, my existing needle needed a firm push to get it into the jet and I couldn't manage to get the conical part to enter all the way into the jet, at all. Checked on a straight edge, the needle did not appear to be bent. It transpired that the needle had been replaced, when the car was in America. But not the jet. The fit between the needle and the jet is to a very fine tolerance. Fiennes in fact, will only sell the two items as a matched pair at £79.42 inc. vat. My needle and jet were mismatched; sometimes they 'worked' and sometimes the needle jammed. So if you cannot tune your carburettor precisely, check that there is absolutely no friction when the needle enters the jet.

Note when re-fitting the carburettor, to the block. Place the fuel feed pipe in position before offering the carburettor to the block.

The throttle actuating rod, goes between the carburettor alloy air inlet elbow and the starter carburettor control rod. Loosely fit the top right and two bottom nuts that hold the carburettor to the block. This will allow the carburettor to be pulled back and give the clearance necessary to fit the top left nut. Alternatively, note that Fiennes can supply new nuts that are deeper & waisted. They make the job so much easier!

## Aluminium washers: Specifications

*High speed jet:* 13.80 mm OD. 9.50 mm ID. 0.70 mm thick (28 thou.)

*Starting Carburettor top cap:* 28.80 mm OD. 21.00 mm ID. 1.18 mm thick (46 thou.)

*Fuel feed union & float seal:* 18.95 mm OD. 11.13 mm ID. 1.21 mm thick (47.5 thou.)

*High and low speed jet cap and Starting Carburettor jet cap:* 22.15 mm OD. 15.95 mm ID. 1.21 mm thick (47.5 thou.)

*Low speed jet:* 14.40 mm OD. 8.30 mm ID. 0.70 mm thick (28 thou.)

*Float chamber drain & Fuel feed banjo:* (2) 15.78 mm OD. 9.56 mm ID. 0.88 mm thick (35 thou.)

A compression test, or a leak down test, is a convenient way of ascertaining the condition of the top half of the engine. Low cost testers are available from Sykes Pickavent. To use the tester, first remove all six spark plugs. Screw the tester into No1 plug hole (nearest the radiator). Hold the throttle wide open with a brick or similar. Without switching the ignition on, operate the starter for ten seconds. Note the maximum reading on the gauge. Remove the tester and pour about half an egg cup of engine oil down the plug hole. Screw the tester back into the plug hole, repeat the test and note the reading again. This second test, is known as a 'wet test'; the oil temporarily seals the piston rings. If the reading on the gauge is much the same as the first dry test, it indicates that the piston rings are in good condition. Any loss in compression is therefore either past the valves, or possibly via a broken head gasket. If on the other hand, the reading on the gauge on the second test, is 10lbs higher, or more, it indicates a faulty piston ring(s). The actual compression figure you should see on a 20HP or early 20/25 engine in good condition, is in the range 70 - 90lbs. The important issue, is that within this range (or higher), it should be the same +- 5lbs for all cylinders.

The second method of engine diagnosis, is to use a leak down tester. This involves connecting an air line to a plug hole via two gauges - one shows the air pressure applied to the bore, the other one measures the loss. There will always be a small loss, even with a brand new engine. But a difference of 25% or more on the gauges, indicates a top end engine problem.

In use, make sure that the bore being tested, has the piston at TDC (Top Dead Centre) with both valves closed. Turn the engine over slowly on the starting handle to achieve this. Remove the rocker cover and you can watch for the position of both valves closed. At the same time insert a thin stick (I use an artists paint brush) into the cylinder via the plug hole and watch for the stick rising to indicate the piston at TDC.

With the leak down tester in position and being fed about 100psi - use your ears!

*1.      A hissing noise from the carburettor, indicates a faulty inlet valve.*
*2.      Remove the oil filler cap; a hissing from the sump, indicates faulty piston rings or a holed piston.*
*3.      Remove the radiator cap; bubbling coolant indicates a broken cylinder head gasket.*
*4.      Listen at the exhaust pipe outlet; a hissing indicates a faulty exhaust valve.*

Repeat the test for all six cylinders. It's an extremely useful tool, to help in gauging the condition of an engine on a car you might be thinking of buying.

De-sludging, is what the factory used to call a 'bottom end de-coke'. It was required, because oil technology and filtration, when these cars were new, was still rudimentary. Once done and with regular oil changes using modern oils, it shouldn't be required again, until the engine needs an overhaul. The phenomena occurred, because the by-products of combustion were carried through the engine, by the engine oil. When the oil was returned to the sump, the heavier elements of this detritus sank and collected in the bottom of the sump. The gauze 'filter' on the oil pump inlet, is next to useless in this context. It will stop a circulating piece of broken split pin, but 99% of the rubbish goes straight on through. On the ends of the six connecting rods, are 'sludge traps'. This was Henry's answer to the problem of removing the debris in the oil. The connecting rod describes a circle as it whirls around; the 'G' force generated by this rotation is truly massive. The G force, flings the particles in suspension in the oil, into the sludge traps, where it collects and solidifies. As the mileage ramps up, so does the sludge. If left in situ, it could eventually block an oil-way and run a bearing. Hence the need for a regular bottom end 'de-coke'. A very large part of the muck is carbon based and measured in microns. If left circulating in the oil, the particles are small enough to remain in the oil film between two bearing surfaces. In any case, the carbon is soft enough not to do any damage. A modern oil, like Rock semi-synthetic 20/50, holds much of this micron sized debris in suspension and is removed, when the oil is changed. Hence the importance of, at least, an annual oil change. However small the annual mileage

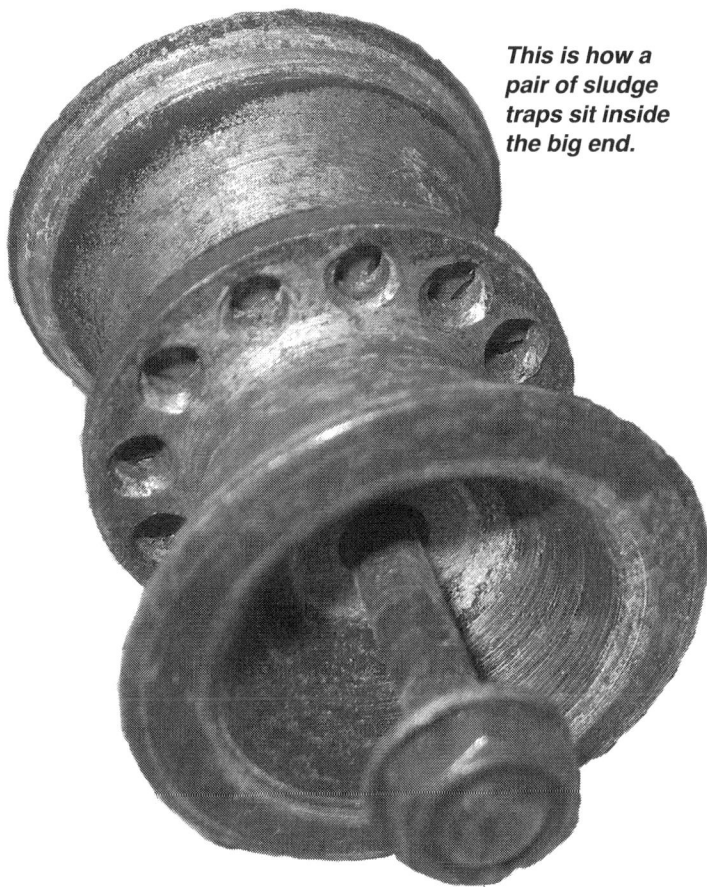

*This is how a pair of sludge traps sit inside the big end.*

might have been. Bigger pieces of debris will collect in the bottom of the sump; the lager part of which, will be removed when the oil is changed. The clue that your engine might need de-sludging, is to watch the oil pressure gauge, when switching off a hot engine. Typically at slow idle (hot) you should see between 5 and 15lbs on the gauge. As the engine comes to rest, this reading should decay to zero quite quickly. If it takes over 5 seconds for your oil pressure gauge to zero, then it is probably caused by narrowed (sludged) oil-ways.

De-sludging is an absolutely filthy job! Drain the engine oil when hot and remove the sump. This procedure alone will take you half a day. Start by removing the starter motor and the bolts that hold the lower half of the clutch cover to the sump.

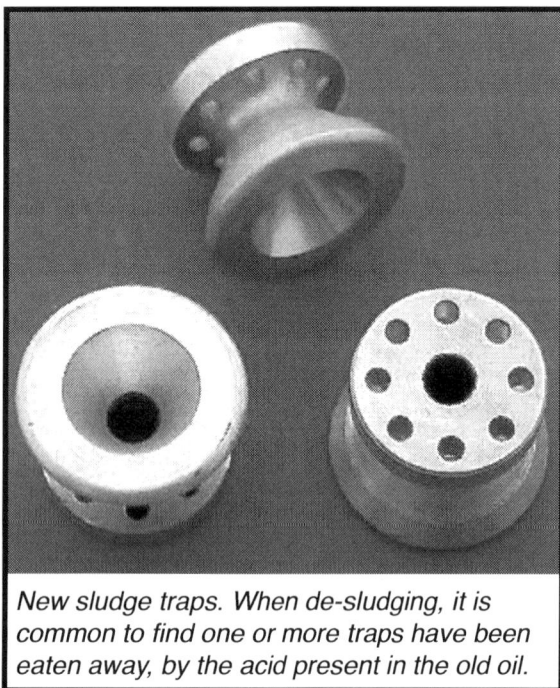

*New sludge traps. When de-sludging, it is common to find one or more traps have been eaten away, by the acid present in the old oil.*

Most of the bolts holding the sump, are hand reamed, so keep them in order and put them back into the same bolt holes. Note there are four studs that stay in the crankcase. Also note, that there are two long studs/nuts on top of the bell housing, that go through the sump. These are actually rods, with a thread at each end. The top thread is peened over and so withdraws from the top - this might mean removing parts from the bulkhead, to give room for withdrawal. However, check first, as sometimes they have been fitted with the peened nut at the bottom. You will find the whole job is much easier, if the radiator is also removed. Plus if the engine needs de-sludging, then chances are, the radiator and block water ways need rodding and pressure washing anyway. There is no need to disturb the front wheel chest. Cut the wheel chest gasket cleanly with a knife where it abuts the sump and make a new part gasket, ensuring plenty of Gasket Sealant on the cut joint when re-assembling.

The amount of sludge collected in the removed sump will amaze you. Clean it all out, using paraffin or white spirit. Next on the cleaning list, are the sludge traps themselves. There is no need to disturb the big-end shells when de-sludging and little point in doing so. Though No 2 and No 5 main bearings do have to be removed to get access to the big-end journals. The main bearings do not have split-pinned nuts. They have plain nuts with bent over tab washers featuring a two point location. Mark the nuts and washers before removal, so that you can replace and tighten exactly as they were.

When the crank is turned to remove a big-end trap, make sure the feed hole to the end, is at the side and almost at the top. Then when the split pinned nut and bolt holding the trap has been removed, the traps can be removed. Use an old feeler gauge, to carefully remove the build up of sludge at the bottom. Remove big blobs of sludge with your finger and then wash out vigorously with cellulose thinners. Irrigate with more thinners, while at the same time use an Aqua vacuum with a 1/4" tube attached, to suck out any left over debris. Make sure the feed hole is completely clean and clear. It is entirely possible to find on a neglected engine, that some sludge traps have been partially eaten away, by the action of the acid within the degraded oil.

Finally, when all the cleaned traps are back in place, inject clean oil into the main oil feed pipe. Place a clean rag over the pipe and air pressurise at 15-20psi. Then get an assistant to turn the crank by the handle, and observe the oil bleeding from all journals; air/oil loss should be uniform from all the journals, bleeding out from the sides of all bearing surfaces. This procedure

confirms oil feeds to all journals and will blow out any particles that might possibly have got into the oil feed drilling. The only disadvantage is you have to lie underneath the engine to inspect all the joints and therefore you'll get covered with oil. When the sump is re-fitted and filled with oil, use the starter motor to spin the engine (plugs out) until oil pressure is registering on the gauge, before starting the engine.

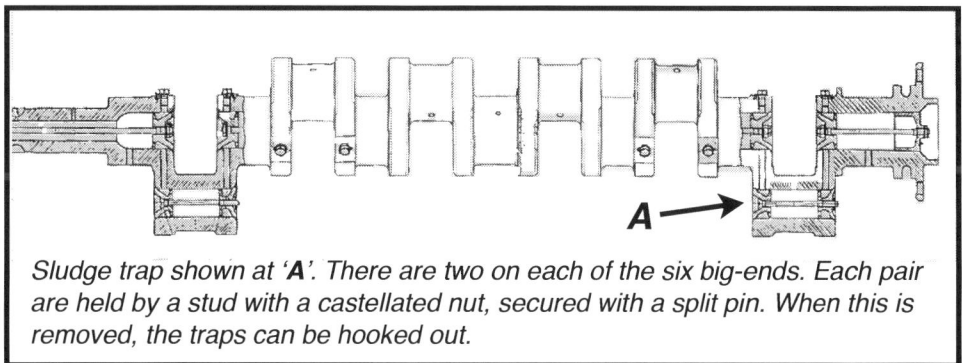

Sludge trap shown at '**A**'. There are two on each of the six big-ends. Each pair are held by a stud with a castellated nut, secured with a split pin. When this is removed, the traps can be hooked out.

When changing the engine oil, if you leave the car for a period of time to completely drain the old oil, an air lock can develop in the oil pressure feed pipe. This will lead to zero oil pressure being shown on the gauge. There is a gland nut underneath the carburettor; start the engine, slacken this gland until oil flows, re-tighten and pressure will once again be registered on the gauge.

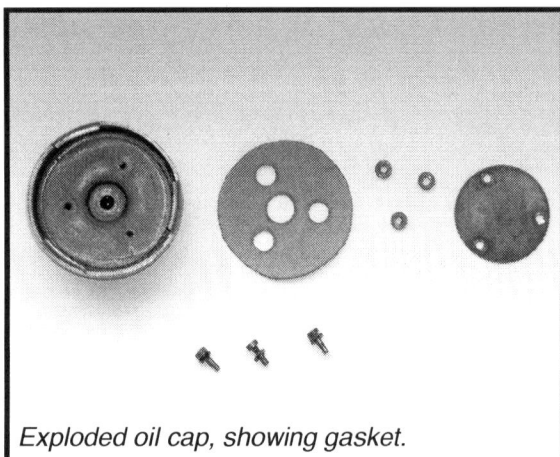

*Exploded oil cap, showing gasket.*

Be aware, that this quirk can also manifest itself, if you have a low oil level in the sump. Enthusiastic cornering, or braking hard, can surge the oil in the sump and lift the oil pick-up pipe. So if the oil gauge suddenly falls to zero, don't panic! Bleed the pipe as before at the bottom of the carburettor, to restore oil pressure on the gauge. Note that when the gauge showed zero, oil pressure to the engine was in fact unaffected. A smell of oil inside the car and/ or excessive oil mist on the engine, are both signs that the oil filler cap is faulty. There is a sealing gasket inside the cap, designed to contain the pressure that builds up in the sump. Cut a new one from thin cork sheet, or thick gasket paper - see the picture on this page.

*Bluebell (GXO71) en route to Anso in Spain.*

42

If you need to replace the exhaust system, you would be wise to fit a stainless steel system. Some owners maintain that stainless 'alters' the exhaust note. I can't say that I have ever noticed this. I have noticed that the stainless steel exhaust system I fitted over seven years ago, still looks brand new. Note that not all 'stainless' systems are the same. Exhaust systems should only be

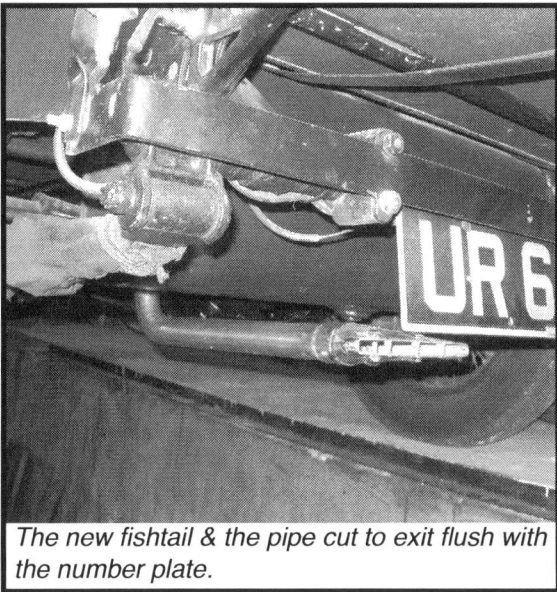

*The new fishtail & the pipe cut to exit flush with the number plate.*

fabricated from low carbon type 304 austenitic stainless steel. Other grades of stainless are not suitable. My system was fabricated by R J Lonsdale in North Yorks at a total cost of £455.31.

Because the Rolls-Royce chassis was hand made, there are tiny dimensional differences which means an off-the-shelf exhaust needs to be fettled. The system delivered by Lonsdale, comes with the correct eight stud connecting flanges and

the rear mounting bracket and fishtail supplied loose. The system was fitted to the car and the position of the flanges, bracket and fishtail marked. I then removed the system again and the loose parts were TIG welded in position by a local shop (£20 cash), before being refitted to the car.

I cut the rear pipe slightly oversize to the factory drawing, so that the fishtail now terminates flush with the number plate and is less likely to soot the bottom of the fitted rear trunk.

Asbestos rope was originally specified to be bound around the exhaust pipe, where it runs close to the rear seat pan, to avoid possible scorching of the wooden chassis closing panel. A modern heat resistant rope seal of the type used on wood stoves, was sourced from an ironmongers and bound around the curve on the new stainless steel exhaust pipe.

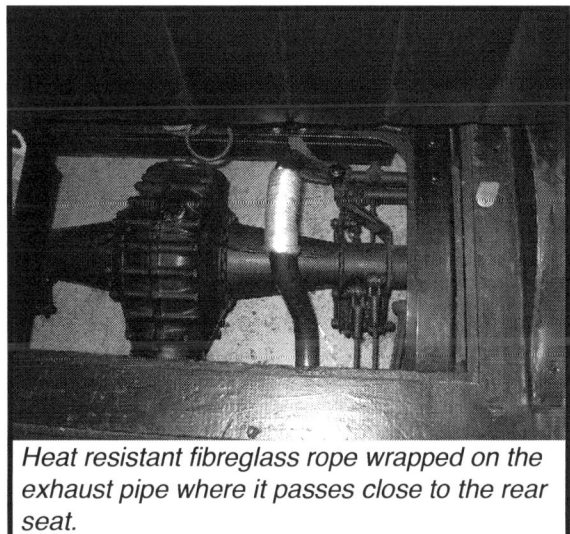

*Heat resistant fibreglass rope wrapped on the exhaust pipe where it passes close to the rear seat.*

The oil pump rarely gives trouble. But if your engine oil pressure is outside the range (hot) of 10 - 15lbs at slow idle and 20 - 25lbs running at 30mph and the bearings are known to be in a satisfactory condition, then it might be the pump at fault. It's easily accessible from underneath the car and should be removed to the bench.

Inspect the top plate (N) for signs of scoring. Remove the marks by lapping both parts of the oil pump casing with fine grinding paste, on a piece of plate glass.

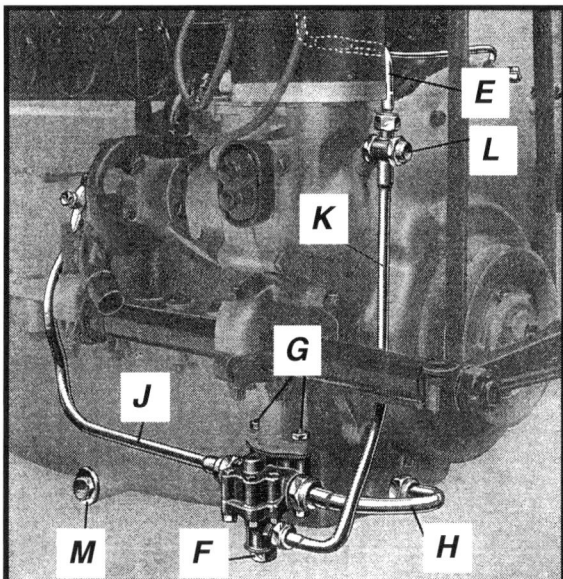

Wash very thoroughly with paraffin. If in doubt as to condition, renew the relief valve spring (circa £10 from Fiennes).

Note the presence of washers on the relief valve. Each washer is .05" thou thick and are designed to increase the oil pressure. If your (hot) oil pressure is low and you have less than the maximum permitted of five washers fitted, then fit another to raise the pressure. Do not exceed 5 washers! Though it has to be said, that engines have been found on strip-down to contain extra washers. Unscrupulous vendors have been known to use this dodge, to disguise low oil pressure on engines overdue a re-build.

Fit new spring washers and gaskets when re-building and clean out internally, the external oil pipes.

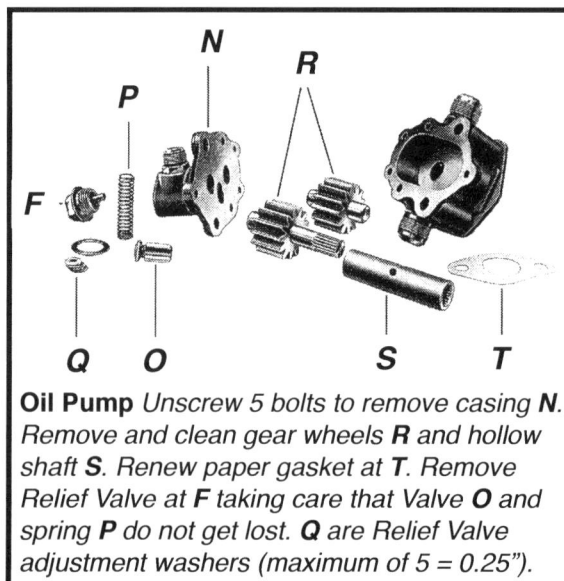

**Oil Pump & Relief Valve:** *Oil is sucked in from the filter by pipe **H** and delivered by pipe **J**. A connection from **J** goes to the dashboard gauge. Pipe **K** delivers overflow from the pump to the timing gears. **L** sprays oil onto the gear teeth. Pipe **E** delivers oil to the cylinder head rocker shaft. Release upper end of pipe **J** to bleed air from the oil system. Plug **F** is the Relief Valve. Undo 2 nuts **G** (& unions) to remove the pump.*

**Oil Pump** *Unscrew 5 bolts to remove casing **N**. Remove and clean gear wheels **R** and hollow shaft **S**. Renew paper gasket at **T**. Remove Relief Valve at **F** taking care that Valve **O** and spring **P** do not get lost. **Q** are Relief Valve adjustment washers (maximum of 5 = 0.25").*

*Note: illustrations are of a 25/30. There are detail differences with the earlier cars.*

The cooling system on the 20HP & 20/25 is more complex than the average car, in as much as swaged copper tubes run through the block and encase the studs, that hold the cylinder head. The main problem that is encountered, is silting up of either the radiator, the engine waterways, or both.

Drain the radiator, remove the top & bottom hose and flush the radiator through, using a garden hose. Do the same with the water pipe connections on the engine. Big lumps of scale visible in the radiator, can often be removed using a magnet. But if the radiator is badly scaled and the engine has a history of overheating, then the radiator core may well need replacing. Or at least, dismantled by a professional and physically rodded through. A radiator re-core is not a cheap job. But be aware that *all* of the usual RR specialists, do not have the facilities to re-

core themselves. A large part of your final bill, will be for removing the radiator and shutters and paying for the standing overhead, while the car sits part dismantled in their premises. If you remove the radiator & shutters yourself, then you can do the whole job for circa £1200. There are at least four specialist radiator shops in the UK; I use Bryan & Son in Tunbridge Wells.

Removing the radiator is fairly easy. On the earlier cars, there are two studs either side of the radiator shell, that go down through the chassis. On the later models, there are two fastenings on a cross

*Shutter removal. A fastening at the foot of each shutter blade is removed and then the blade with its peg, is removed from the top operating link. In this picture, four have been removed from one side.*

*One of 2 fastenings under the front axle, that holds the radiator. The picture is of a 25/30, but the other models are similar. Note the position of washers, collars, springs, when you remove.*

member above the front axle, plus one at the top of the radiator holding the stay to the bulkhead. The shutter control also needs disconnecting. The studs are usually secured with a split pinned castellated nut. When re-assembling, these nuts should be tightened until they 'bottom', then backed off to the nearest hole, to get the split pin through. The arrangement is to hold the radiator firmly, while at the same time, to allow for some chassis flexing.

The radiator is heavy and is a two man job, to lift out. Removing the shutters is self evident - but do take pictures as an aid to replacing. At the top there are two springs that return the shutter leaves; these are often weary and will need replacing.

On the later cars, the shutters are operated automatically by a calorstat, fitted into the header tank. These can prove to be extremely difficult to remove and expensive to replace. If yours no longer works, a better solution is to disconnect the shutters and leave them permanently wide open. Then fit a modern thermostat. A kit is available from Alan Murcott and once fitted, is totally unobtrusive.

If the engine is silted up, then you will need to remove the front and side plates from the block. The 2BA screws holding these plates are likely to be corroded. Clean the screw heads with a scriber, soak with WD40 and then use an impact driver to start the screw. Moderate force only, as there is a risk of cracking the plate if you hit it hard. Then use a long screwdriver with a perfectly fitting tip. I bought a new one with a hexagonal shaft. By leaning really hard on the end and using a spanner on the shaft, all but one screw removed. The one that didn't, I ground away the head of the screw and once the water plate was off, a pair of mole grips undid the remains. Once the six water plates are removed, chip away all the silt and muck you will find inside. The next step, is to remove the many core plugs on the block and cylinder head. There are 17 of these on the later engine! The only way you will remove the core plugs, is to make up a tool. I cut a circle from a piece of 4mm steel sheet and drilled 4 holes to match the holes in the core plugs. To make the pegs, I used an old drill bit that fitted snugly in the holes, cut the shaft into four and Mig welded the pegs into the steel circle. Finally I cut the circle down almost flush with the pegs and then welded the circle/pegs into an old box spanner. To encourage the cores to move, score around the thread and liberally dose with WD40. Give it a few minutes, clean surplus WD40 off and then squirt each plug with Loctite 'Freezer'. This aerosol product rapidly cools the plug, allowing differential expansion to break the seal between the plug and the block. Then using a long lever torque wrench on the box spanner tool, they will come out. When the core plugs were removed, I was surprised to find that beneath almost all, was a wall of what appeared to be graphite. All of this muck needs to be picked out and then the block and head pressure jetted with water, through the core plug holes and the water plate apertures. Muck will pour out in large quantities.

When re-assembling, cut new gaskets for the water plates. If any of the screw threads have stripped, you will find that M5 is just oversize and will tap easily into the damaged 2BA thread hole. Stainless steel M5 screws are readily obtainable and look the same as 2BA once in place. Do not use sealants on the core plugs, as this will make them even more difficult to remove next time. If you do subsequently experience any water leaks, better to pour a couple of cans of Wynns Stop Leak into the radiator. This product will not block the radiator core, or cause silting. Once you have done all this work and achieved a cool running engine, whatever the ambient temperature,

**A**: Core plug removed, revealing that the hole is full of muck. The one on the right has been cleaned out. **B**: Water plate removed and silt cleared out. The cylinder bore can now be seen.

*Front water plate removed. The soft muck has already been cleaned out. Now the chipping and scraping can begin, before the water cavity is pressure jetted with water.*

which 'works' all the other shutters. Lubricate all twenty-eight ball sockets on the shutters with anti-seize copper grease, before re-fitting.

As a matter of course, fit new radiator hoses and a new fan belt (Part No B41). The most economical you will not want to do it again! The best prevention is to flush out and refill the radiator *annually*. And to use a 50/50 mix of anti-freeze and ionised or distilled water. Note that this high strength mixture, is more about the corrosion inhibitors that are in anti-freeze, than in the protection from extremely low temperatures. *Note that the anti-freeze needs to have IAT and **not** OAT inhibiters.* 99% of the anti-freeze product suitable for the older car, is blue in colour. The long lasting modern variety that will degrade the gaskets in our cars, is usually a brown or orange colour.

When re-building the shutters back into the radiator, note that the shutter either side of the badge, carries an extra crank and

*This new housing contains a modern thermostat. Much more efficient than the calorstat in the header tank. This is left in place, but the link connecting the calorstat is unfastened and the radiator shutters left permanently open.*

way of buying hose, is to purchase one
metre lengths from hoseworld. Finally, fit
new bonnet tape to the radiator perimeter.
This is available by the metre from Woolies.

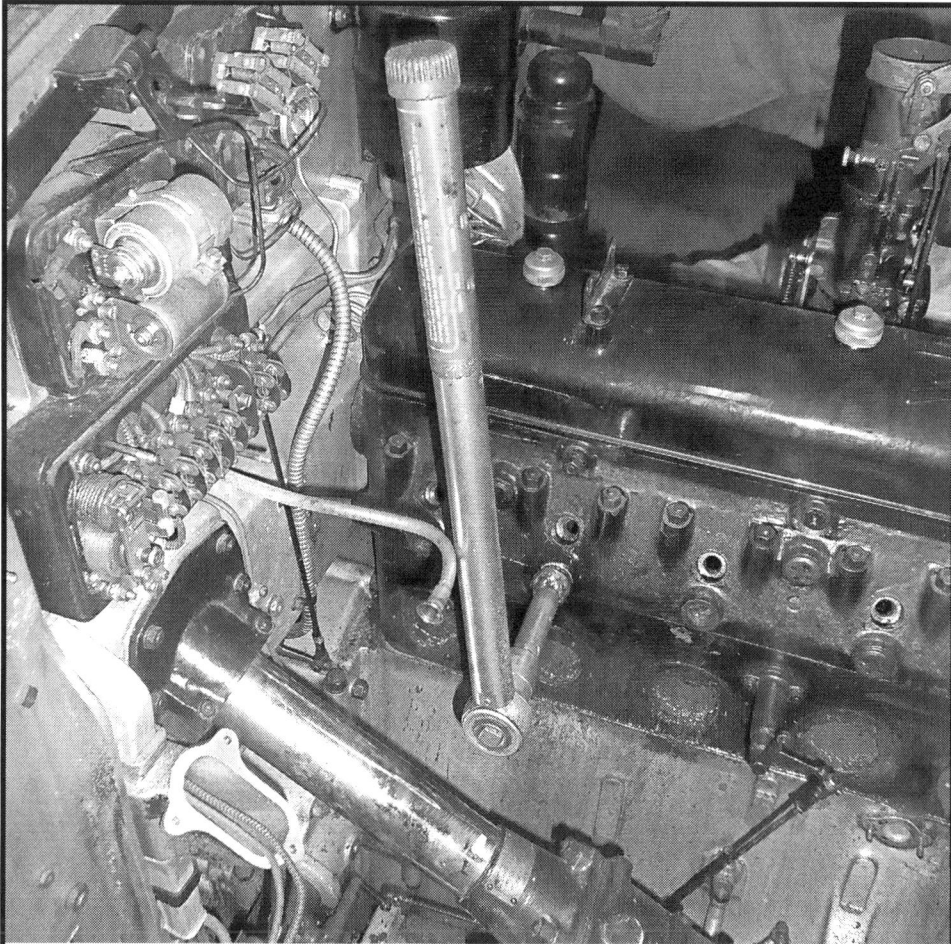

*Homemade tool, removing one of the many core plugs.*

The water pump is not prone to wear, provided that the greasing cup is filled with the correct water pump grease. The cup should be tightened by hand until light resistance is felt. The factory recommended this is repeated every 2,000 miles; I do mine every 500 miles. When greasing - do not not over tighten! All you will do, is force the grease past the seals and splatter it all over the engine.

On Bluebell, there was a persistent coolant leak from the pump, so I removed the pump and dismantled it on the bench. The stainless steel impeller shaft was badly scored, consistent with faulty lubrication and the thrust button had more or less disappeared. This had allowed the impeller to contact the body and which was also scored. Inspection revealed that the lubrication channel that should be visible underneath the grease cup was absent. The conclusion is that the pump was re-built incorrectly when the engine was overhauled in America. The pump dismantling coincided with a visit to Fiennes Restoration in

*The water pump with new parts waiting to be fitted (minus the gland packing pieces, that fit either side of the spring, on this early pump). The inset shows the order of assembly of the various bushes. At the bottom is the modern polythene replacement thrust button (the original was made of lignum vitae wood). The thrust button fits between the end of the impeller and the bronze body of the pump itself.*

Oxford, to attend a Restoration Seminar; so I took the pump with me, to purchase parts. According to Will Fiennes, the shaft and/or impeller was of the wrong type. As this would take some time to sort out, the pump was left with them for overhaul. Fiennes completely rebuilt the pump with all new bushes, impeller and stainless steel shaft. The shaft had to be machined to size before fitting. Total cost including VAT was £467.29.

Had the original components within the pump been correct, then overhauling it with new bushes, thrust button and packing would be a DIY job for a reasonably competent amateur mechanic. Before re-building, strip any remaining paint from the casing and re-enamel the pump gloss black, as per the original finish. If fitting a new thrust button, it will probably need easing with fine wet & dry to get it to fit. Note that the button should be introduced square to its locating hole, and then pressed into place by bolting the two halves of the water pump together. This ensures that the end float on the shaft is zero.

Underneath the knurled nut on the forward end of the pump is a gland packing. This allows the shaft to revolve, while at the same time providing a waterproof seal. This packing was originally Russian tallow. Fiennes supply a modern material that is easier to fit and is longer lasting. There are slightly different arrangements as to how this packing material is contained. Observe and make notes as you take the pump apart.

Fiennes supply detailed instructions as to how to fit their packing pieces. Finally, do not over tighten the knurled ring on the forward end of the pump. It should be hand tight; anything more and you will accelerate wear on the packing.

The Autovac is a vacuum operated fuel pump, situated on the bulkhead under the nearside bonnet panel. It lifts fuel from the petrol tank to the small tank that is part of the Autovac. From this small tank, the petrol runs by gravity to the bulkhead fuel filter and on to the carburettor. The unit is very dependable as there is little to go wrong. So robust in fact, that Autovac's were fitted to London buses right up to the early 1950s. If it is not known when the Autovac was last serviced, then it would be advisable to remove the unit from the car and strip it down on the bench. It is self evident how it comes apart. Thoroughly clean all the parts and check that the float is in good condition. Refer to the Autovac workshop manual on the following pages.

Annual servicing consists of cleaning or renewing the thimble sized gauze filter that will be revealed when pipe 'C' is disconnected at the top of the Autovac. Next remove the screw shown by the arrow and collect the clamp plate that is underneath. Push a screwdriver or a large nail into the pipe and gently twist, to loosen the 90 degree taper connection. This will now be able to be lifted out of the top of the Autovac. Inside the elbow, will be found the tiny gauze filter. Undo the thumb screw shown at 'K'. Be careful as this is removed, as there is a small spring inside. Remove the spring and piston cup that will be revealed.

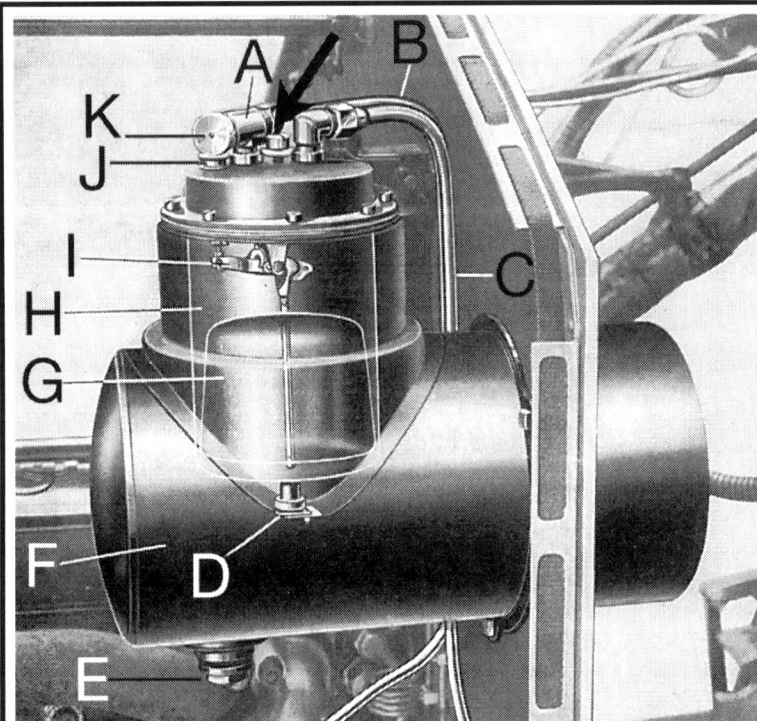

*Autovac Petrol Pump 20/25*

*A: Piston/spring on vacuum side that controls tick-over.*
*B: Vacuum pipe from inlet manifold.*
*C: Petrol feed pipe from petrol tank.*
*D: Drop valve to deliver fuel from vacuum chamber.*
*E: Drain plug (to remove sediment).*
*F: Main tank from which petrol flows by gravity to carb.*
*G: Float that operates valves to control fuel flow.*
*H: Delivery tank in which vacuum is created.*
*I: Spring loaded toggle operating petrol & vacuum valves.*
*J: Air vent controlled by vacuum valve I.*
*K: Pinhole regulating vacuum at tick-over.*

Inside is a felt disk. Renew this by cutting one from a piece of table top felt, available at 'House of Sewing' or similar. This felt regulates the vacuum. When it gets blocked, it can result in a noise that sounds like an angry wasp. If renewing the felt does not stop the noise, then use two felts; or one felt and a disk of paper. *Note that not all Autovac's are fitted with this piston cup.*

A complete overhaul kit containing gaskets, filter etc is available from the Autovac specialist in the UK for about £40. This is Martin Hull in Oakhampton.

If you do have an Autovac problem on the road, then bear in mind that the header tank that is part of the Autovac holds about a litre of petrol. So it is possible to remove the petrol feed (C) and to fill the Autovac from a can/ funnel. This will give you about 4 miles of running, before you have to stop and refill the Autovac again. Tiresome, but it might get you home in an emergency!

I had an obscure problem with my Autovac; it is extremely unlikely this would be encountered on another car, but I include it, to demonstrate that 'difficult' Autovac problems, usually have simple explanations.

The problem was zero fuel in the carburettor

bowl. The car was limped home, by pouring petrol from a can into the Autovac inlet every couple of miles. Subsequently a piece of transparent plastic tubing (to observe fuel flow) was pushed into the inlet of the Autovac and the other end clamped with a Jubilee Clip to the petrol pipe coming from the petrol tank. With the engine running, there was no fuel delivery, until I put my finger over the tiny air hole on the Autovac suction elbow. Fuel then flowed into the Autovac, surprisingly strongly. I renewed the felt, piston and spring within the (cleaned) suction elbow - but the problem remained. The Autovac was then sent to Martin Hull, who operated it on his test rig.

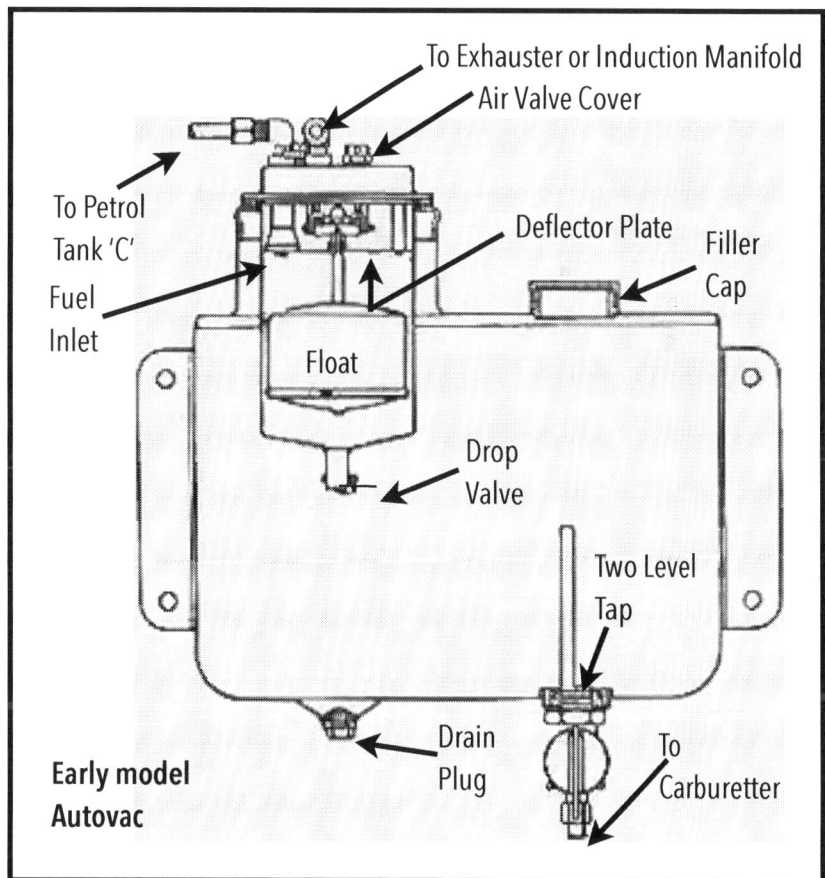

To Exhauster or Induction Manifold
Air Valve Cover
To Petrol Tank 'C'
Fuel Inlet
Deflector Plate
Filler Cap
Float
Drop Valve
Two Level Tap
Drain Plug
To Carburetter

**Early model Autovac**

He reported there was nothing wrong with the piston and spring. The problem was the petrol inlet elbow. This had been re-nickeled and Martin stated that the plater had removed too much metal from one side when polishing. The tapered bore was no longer perfectly circular and the Autovac vacuum was being partially destroyed, by the ingress of incoming air. This would tend to 'suck' the petrol and so there would be no outward sign of a petrol leak.

When refitting the Autovac, note that there is an indent on the top plate. Two cork gaskets are used for the top cover. One above and one below the flange of the inner tank. The indent on the flange of the outer tank must register with the correct hole in the top cover,

The following is the official workshop manual from the 1920s. The Autovac pictured is an earlier model as fitted to the early Rolls-Royce 20HP - but the principals are exactly the same for later models.

### DESCRIPTION

*The Autovac ls divided into two chambers, the inner or vacuum chamber 23 being connected to the exhauster 2 and main fuel tank 3 and the lower or outer tank to the engine supply.*

*Communication between the two tanks is by the drop valve 24. The suction from the exhauster creates a partial vacuum in the inner tank 23, thus closing the drop valve and drawing up fuel from the main tank. As the fuel flows in, the Float 22 rises and operates two valves. One valve 14 cuts off the suction, the other 13 admits air. The admission of air destroys the vacuum,*

*releases the drop valve and allows the fuel to flow into the outer tank. The outer tank is always open to the atmosphere through the air vent 9 and the fuel feeds the engine by gravity. The Autovac is fitted with a Patent Self draining Float 22. The Float Stem is hollow, and has two holes drilled through it, one inside the Float and one outside. If fuel enters the Float through a leak, it is automatically evacuated through the stem during the suction period. During the period of atmospheric pressure, air flows into the body of the Float, enabling it to function as when air-tight.*

### MAINTENANCE

*Remove foreign matter in Filter 4 and sediment trap at base of outer tank. To remove Filter, disconnect fuel supply pipe from elbow 3 and slack off clamp screw 7. Sediment from the fuel collected at the base of the outer tank can be drawn off by removing the Hexagon Plug.*

*The pipe connections 5 must be tight and free from leakage. The screws 11 holding down the top cover must be tight. The elbows must not he reversed, the one with the Internal valve 2 is the suction elbow.*

*Clean any extra filters in the supply system. The valves 13 and 14 are pinned and sweated into position and require no subsequent adjustment.*

### SHOULD ENGINE STOP & AUTOVAC IS SUSPECTED

*Turn off Petrol Tap and disconnect the engine supply pipe at Autovac. Next open Tap to see if you get a flow of fuel, making*

1. Top Cover c/w deflector plate and supports.
2. Suction elbow c/w nut & nipple for 3/8" OD pipe.
3. Fuel elbow c/w nut & nipple for 3/8" OD pipe.
4. Gauze filter cone.
5. Hexagon Union Nut for 3/8" pipe.
6. Nipple for 3/8" pipe.
7. Clamp screw.
8. Clamp Plate.
9. Domed Air Vent Cover.
10. Spring washer for Clamp.
11. Top Cover Screw.
12. Cork Washer for Top Cover.
13. Air Valve with Rocker and Pivot.
14. Suction Valve with Rocker and Pivot.
15. Valve Lever.
16. Float Toggle Lever.
17. Screw & Nut for Valve Lever.
18. Valve Pivot 7/16" long.
19. Float Pivot 1/2" long.
20. Valve Spring.
21. Float Link.
22. Float complete with Stem.
23. Inner Tank with Drop Valve.
24. Drop Valve Complete.

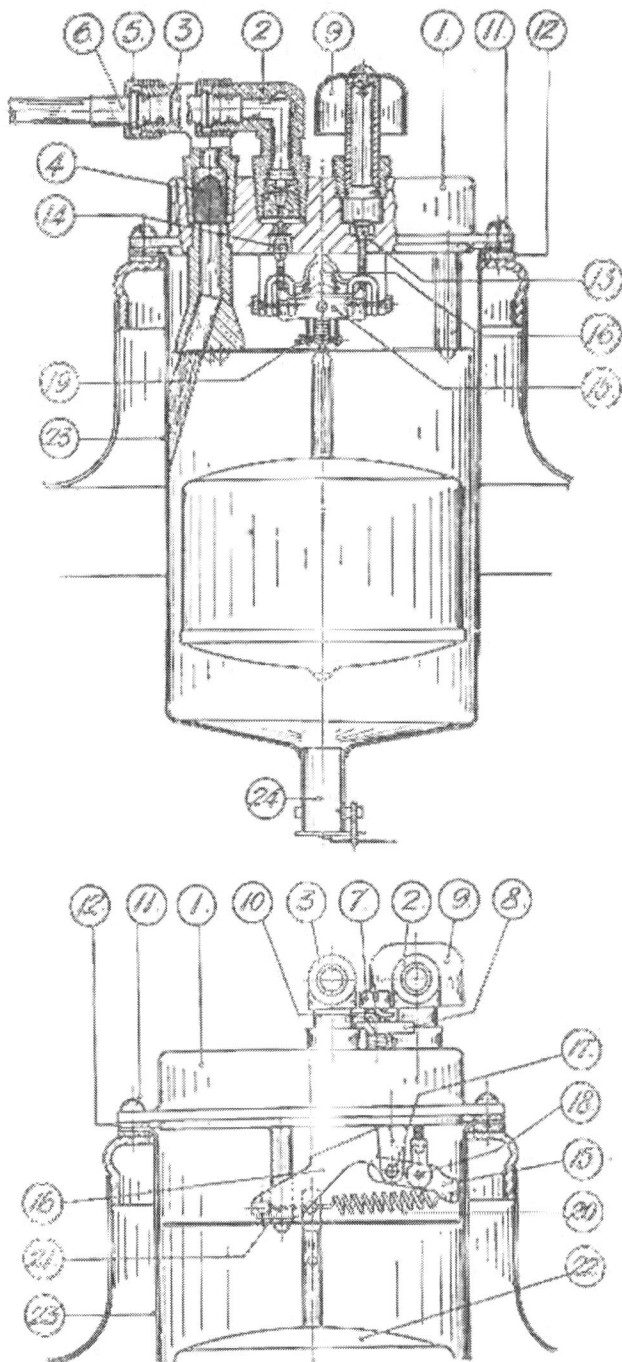

sure at the same time that the Tap is not choked with foreign matter.

If Autovac is found to be empty then it, or the connecting pipes from exhauster or supply, can be suspected. Disconnect main fuel pipe at Autovac and try to lift fuel by sucking on this pipe (blowing is a false test). If fuel cannot be lifted by this method, then

55

examine pipe for cracks or restrictions. If no faults are found in the main fuel pipe, then make a test on the suction pipe, by disconnecting at Autovac (this will have to be made by running the engine, in which case a small supply of fuel will have to be put into the Autovac, so that the engine can be fed by gravity while running). When engine ls running, test for suction by placing your finger over the pipe. If no pull ls felt, then suction pipe should be examined for cracks or restrictions.

If after the pipes have been tested and nothing is found wrong, then the Autovac can be suspected, in which case we recommend a replacement or the apparatus in question taken to a service depot for overhaul. We find that the chief cause of failure is accumulated wear, but only after very long service and our remedy is to change all the valve mechanism with the exception of the float.

Replace spring 20 if stretched or broken. To test drop valve, remove inner chamber and immerse drop valve in fuel, making sure face of disc is clean before doing so. If fuel enters the tube fairly quickly, valve ls leaking and requires renewing.

*When the Autovac is being reassembled - take careful note of the following instructions*.

Two re-enforced cork packing washers 12 are used for the top cover. One above and one below the flange of the inner tank. Air leakage into the inner tank will reduce the suction. The Small Vent pipe projecting from the flange of the outer tank must register with the correct hole in the top cover, which connects with the air vent 9. The link 21 holding the float to the toggle lever 16 and the spring 20 must be looped into the top hole in the float stem. If it should be necessary to prime the Autovac by hand, pour in about one pint of fuel - if the Autovac has no filler to the outer tank, remove fuel elbow and prime through fuel inlet.

This really is belt & braces! But if you intend long distance touring in your car, you might feel the need to carry an emergency fuel pump in the toolbox. The pump itself, is a standard 12volt SU bulkhead pump. Available new from Burlen Fuel Services. You'll also find used examples on eBay. Make up a sheet metal holding bracket as per the picture. The slot measurements are such, that the bracket will fit underneath the thumb screws that hold the bulkhead fuse box on the 20/25. Make up wiring as shown in the picture. The +12v cable to the pump, screws under the right hand thumb screw on the bulkhead mounted ballast resistor. The negative cable from the pump terminates in a bulldog clip and clips onto any bare metal on the bulkhead. A rubber fuel line then goes onto the INLET connection on the pump. The other end goes onto the pipe leading to the petrol tank, disconnected at the top of the Autovac. It will push over the nipple on the pipe and both ends of the pipe, are secured with Jubilee clips.

Two short lengths of rubber fuel pipe are cut and connected to a fuel pressure reducer as shown in the picture. The reducer is also available from Burlen Fuel Systems. It is necessary, as the fuel pressure to the Rolls-Royce carburettor is designed to be 3lbs. The electric pump outputs higher than this. If not reduced, it will result in fuel constantly dripping from the carburettor float chamber. The pipe with the Reducer, connects to the metal pipe from the bulkhead fuel filter feeding the carburettor. The other end connects to the OUTLET on the SU fuel pump. You will also have to

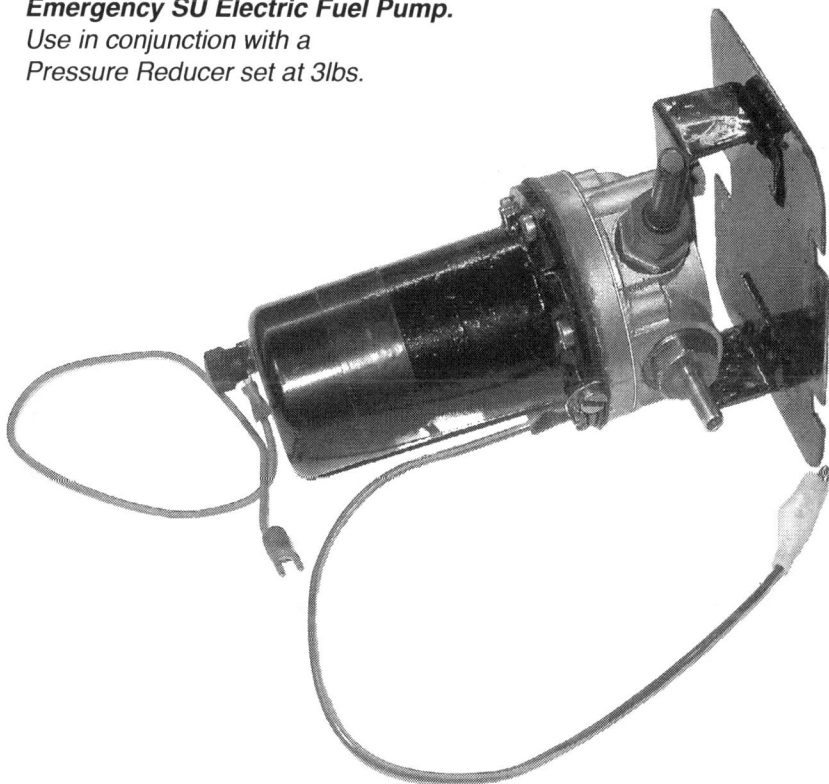

**Emergency SU Electric Fuel Pump.**
Use in conjunction with a
Pressure Reducer set at 3lbs.

*Emergency electric fuel pump in position. +12v cable yet to be attached. The cable screws under the thumb screw arrowed in white.*

buy, a spare pipe connector which normally fits on the fuel filter, to get a leak free seal on the emergency pipe connection to the carburettor. Fiennes can supply this.

All emergency fuel lines are in rubber and held by jubilee clips.

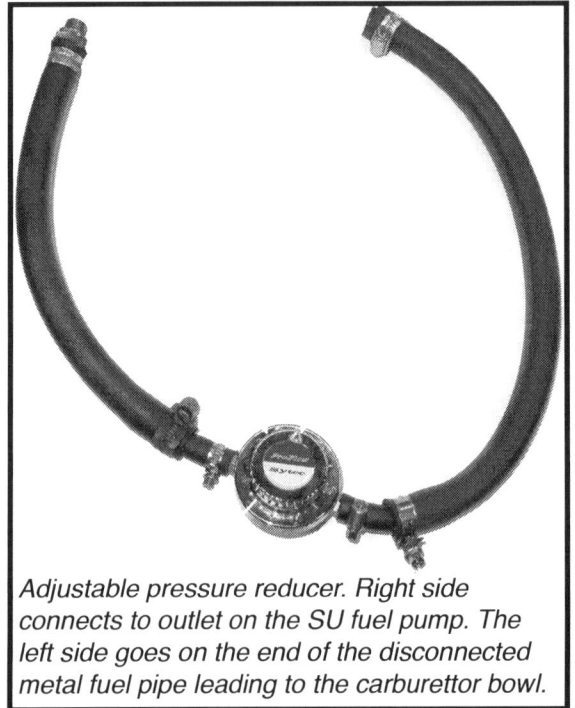

*Adjustable pressure reducer. Right side connects to outlet on the SU fuel pump. The left side goes on the end of the disconnected metal fuel pipe leading to the carburettor bowl.*

There are two primary fuel filters, plus a small gauze third one, inside the Autovac inlet elbow. The primary filter inside the petrol tank, has a very large surface area and shouldn't require attention outside of a major overhaul. Which is just as well, as some coach builder's paid scant attention to providing access, to the top of the petrol tank, when they built the body. The main fuel filter that does need regular cleaning, is mounted on the engine side of the front bulkhead, underneath the steering column. This consists of an aluminium canister, held by a bottom mounted thumbscrew and containing a metal filter element. This element is difficult to clean properly. Some owners swear by lemon juice, others use cellulose thinners. But the fact is, even when spotless, it is a crude filter by modern standards.

Which is why I modified mine, to accept a modern micro filter. This is a standard off-the-shelf Mahlé in-line petrol filter Part No KL13. The outlet of this polythene filter should be cut back to 4mm from the nipple. You will then find that it is a tight push fit into the original outlet hole inside the original Enots filter casing. A coil spring from an old torch placed in the bottom of the casing, ensures that the new filter is not vibrated out of its push fit. Assembled within the original casing on the car, this modern micro filter is completely invisible and will ensure that no muck at all, finds its way into the carburettor.

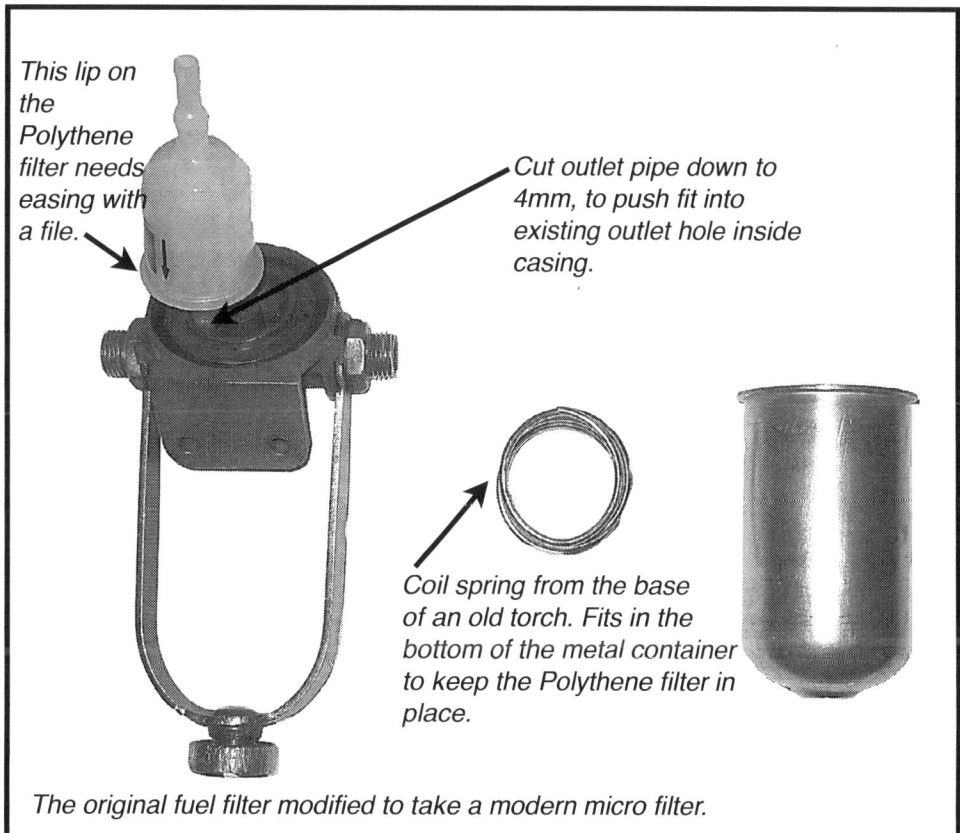

This lip on the Polythene filter needs easing with a file.

Cut outlet pipe down to 4mm, to push fit into existing outlet hole inside casing.

Coil spring from the base of an old torch. Fits in the bottom of the metal container to keep the Polythene filter in place.

*The original fuel filter modified to take a modern micro filter.*

Annually slacken, but do not remove, the drain nut on the bottom of the petrol tank. Dribble about a jam jar out, this will ensure any water in the tank is drained away.

A mystery 'failure to proceed' is sometimes caused by blocked breathers in the petrol tank. If the car is running OK, but the engine stops a few miles down the road after filling with petrol, suspect the breathers! On the earlier cars, the filler cap is screwed onto to the neck of the fuel filler. The last four complete threads are machined off, creating a lip at the bottom of the dished cap. Registering with this lip on the inside of the filler neck, are two vent holes. These holes face the body so are difficult to see at the best of times. The problem of a blocked breather, can be compounded (as it was with my car) if the thread of the petrol filler cap has been greased. Road muck is attracted to the grease; in my case so much so, that the vent holes were completely covered.

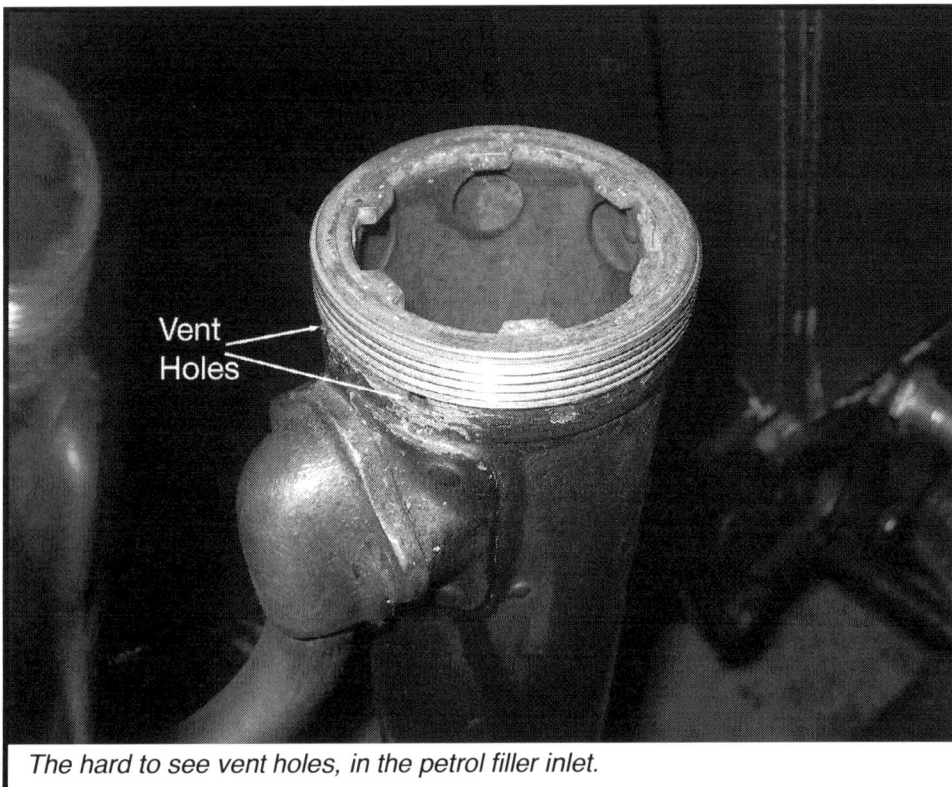

*The hard to see vent holes, in the petrol filler inlet.*

60

On early cars, the fuel tap is located underneath the Autovac. On later cars it is located on the inside bulkhead within reach of the passenger's feet. If you are carrying a passenger in the front, the car is running nicely and then the engine suddenly stops - check that the passenger has not accidentally kicked off the fuel tap! If this is the case, after switching the fuel back on, unscrew the top of the carburettor bowl a few turns, wait for petrol to spurt and then re-tighten, before trying to re-start.

The main problem that occurs with the fuel tap, is degradation of the circular cork gasket that is inside. Ethanol is now being introduced into petrol supplies and it 'attacks' cork over time. This usually results in a leaking fuel tap. The cork gasket, has two holes punched in it and these locate onto two spigots inside the tap. The spigots stop the cork from rotating. However it is often the case that ethanol will eat away at the cork, reducing its thickness, to the point that the cork slips out of the spigots. The gasket will then rotate when the tap is operated, blocking the fuel supply outlet and the engine will stop. A quick check is to ensure the tap is ON and then unscrew the top of the carburettor bowl. If no petrol flows after a minute or so, then suspect the tap. Fiennes was supplying new cork gaskets together with a tube of special gunk, that is supposed to seal the cork against ethanol attack. A better solution in my view, is to cut your own gasket from 4mm thick 'Teflon' sheet. This is impervious to ethanol and is a longer term solution. Whether you use a cork or a Teflon gasket, it is a good idea to permanently stop it rotating within the tap, by applying 2 small blobs of Epoxy paste, to the spigots, to secure it.

Dismantling the tap is self evident. It's easier if you remove the Autovac from the car, as it is so much easier to work on the tap, when it's on the bench. Note that the central castellated nut on the tap is secured with a small split pin. The nut should be tightened until the operation of the tap is really stiff. Then back off to the nearest holes to insert the split pin.

*New fuel tap cork gasket (left) and one removed from the tap after two years service. Ethanol attack has materially reduced the thickness. The two raised pieces that are untouched, is Epoxy resin, that was used to fix the gasket in place, to stop it rotating within the tap.*

For two years, the early cars, up until about May 1932, were fitted with a Hobson petrol gauge. Even from new this was a troublesome instrument. In fact Herbert Austin claimed, that he received more Warranty claims over the Hobson gauge fitted to his Heavy 20 model, than any other component. Hobson liquidated their automotive division soon after. Today, it is rare to see any car that still has a working gauge. But it is possible to restore it.

First a note of caution. The Tetrabromoethane liquid that is used in the gauge is extremely toxic. Handle with care and use rubber gloves at all times. There is only one source of supply in the UK, that I am aware of; Fiennes Restoration.

### How it works

The gauge, utilises the pressure resulting from the height of the fuel in the petrol tank, to activate a column of high specific gravity fluid (SG 2.966), in a glass sight tube, mounted on the instrument panel. The basic principle, is that the pressure exerted by the height of the fuel, is transmitted through a fine bored pipe to the sight gauge. Here, it is balanced by the pressure exerted on a column of Tetrabromoethane fluid. This fluid is almost four and a half times as heavy as petrol. Therefore the change in pressure of [say] a standing height of ten inches of petrol in the fuel tank, will result in circa two and a quarter inches height in the dash sight gauge. The dash unit is basically a 'U' tube manometer with one leg of the 'U' forming the sight gauge and the other leg, which is used to

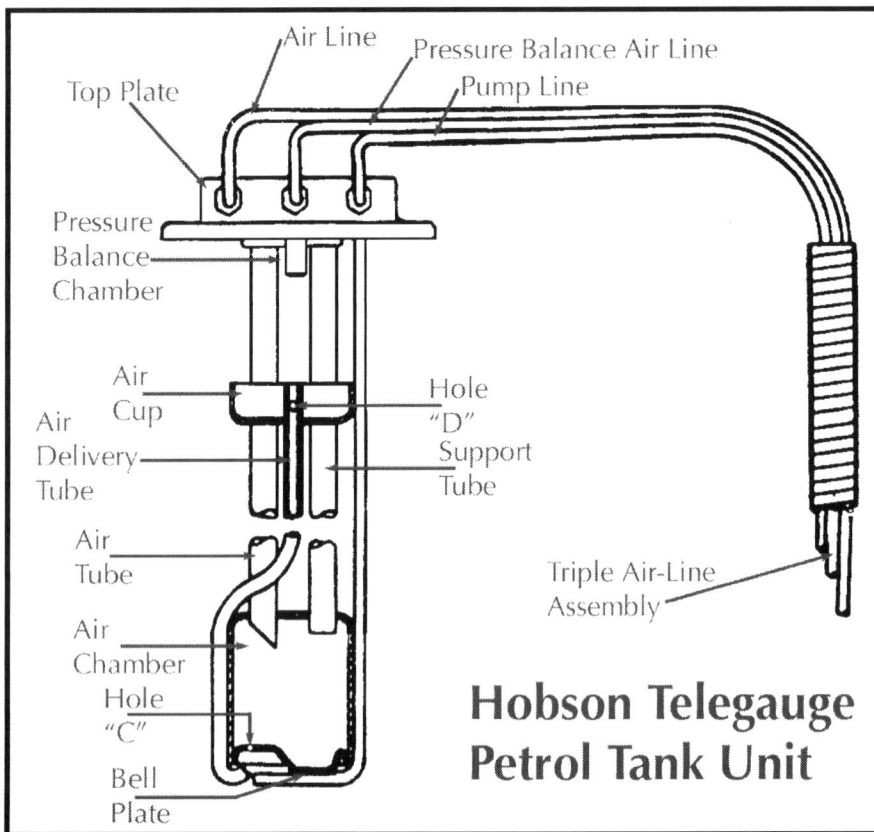

Air Line
Pressure Balance Air Line
Pump Line
Top Plate
Pressure Balance Chamber
Air Cup
Air Delivery Tube
Hole "D"
Support Tube
Air Tube
Triple Air-Line Assembly
Air Chamber
Hole "C"
Bell Plate

**Hobson Telegauge Petrol Tank Unit**

initially adjust the instrument. hidden at the back of the gauge. The glass and the brass tube are connected by a length of copper tube.

## Operation

Pressure from the petrol in the tank is built up in an air chamber situated near the bottom of the petrol tank. Petrol flows into the air chamber through the hole "C". When the car is on the move, petrol and trapped air bubbles flow through the hole "D" and down the air delivery tube. The air bubbles are released at the bottom of the tube under the air chamber, again entering through hole "C" and so displacing any petrol which may be present in the air chamber. When the air chamber is full of air, further air passing down the tube is released back into the petrol tank. To allow for any leakage in the connecting line between petrol tank and gauge and to maintain a constant height of fuel within the air chamber, provision is made to maintain pressure within the air chamber, proportional to the height of fuel in the petrol tank. When the car is at rest, this tube is full of petrol and is at the same level as the petrol within the tank. When the car is moving, petrol is splashed into one of the cups that are formed around the pipe. It then runs down the air delivery tube carrying with it, trapped air in the form of bubbles. The bubbles then rise and are trapped in the air chamber. The pressure balance line is arranged so that it does not allow air to escape, but serves as a vent, if excess pressure from heat or other causes is developed in the petrol tank. If the car is stationery, then three or four strokes on the dash mounted air pump, will simulate the splash of fuel and produce air bubbles at the

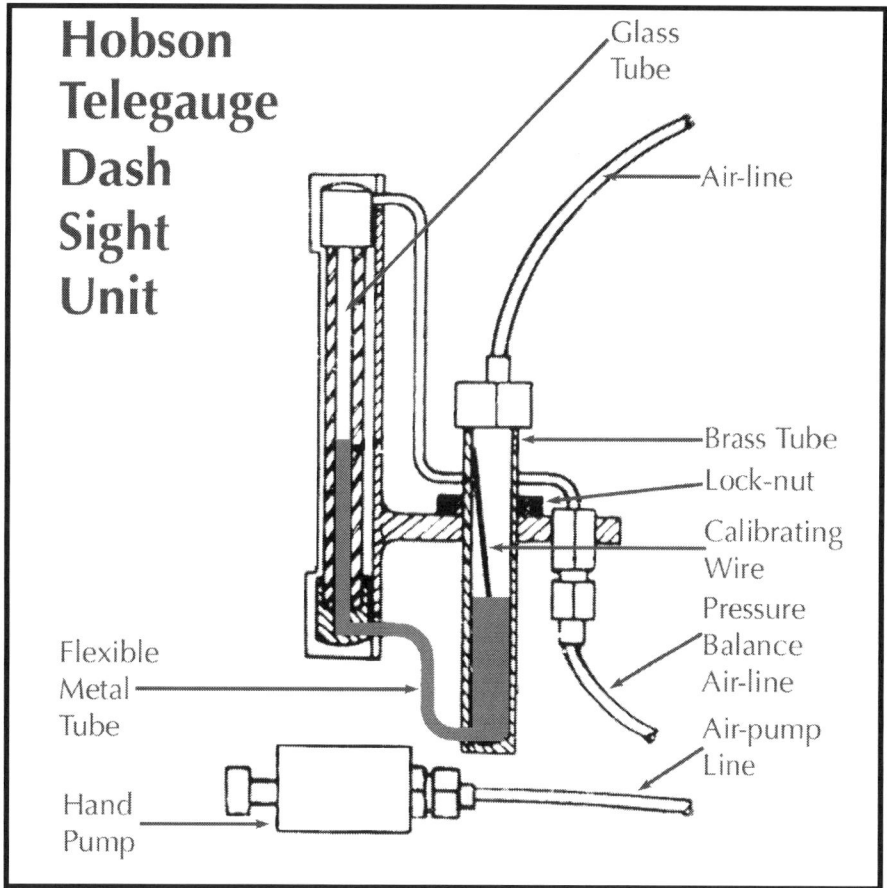

**Hobson Telegauge Dash Sight Unit**

Glass Tube

Air-line

Brass Tube

Lock-nut

Calibrating Wire

Pressure Balance Air-line

Air-pump Line

Flexible Metal Tube

Hand Pump

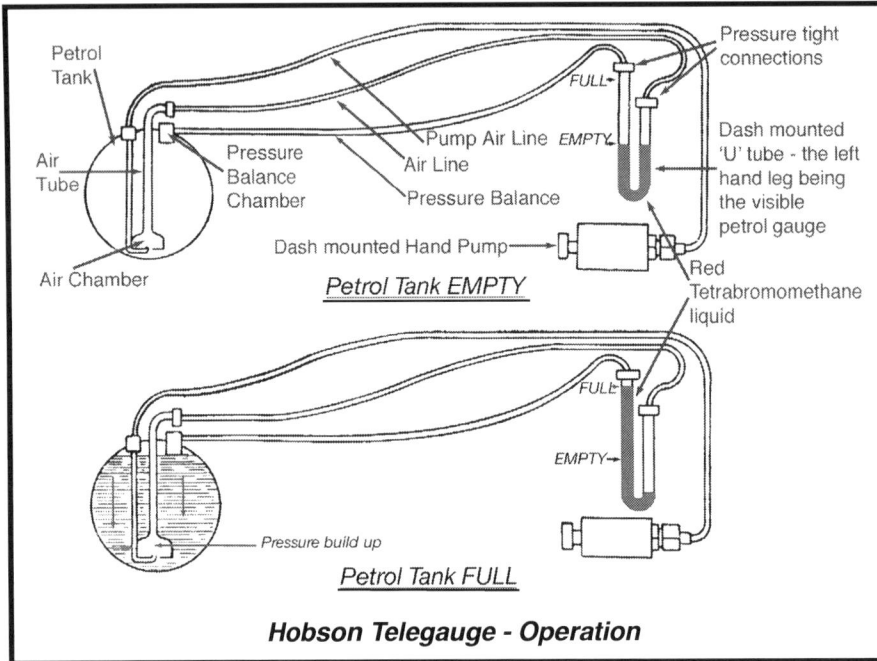

Petrol
Tank

Air
Tube

Pressure
Balance
Chamber

Air Chamber

Pump Air Line   *EMPTY*
Air Line
Pressure Balance

Dash mounted Hand Pump

*FULL*

Pressure tight
connections

Dash mounted
'U' tube - the left
hand leg being
the visible
petrol gauge

Red
Tetrabromomethane
liquid

*Petrol Tank EMPTY*

*FULL*

*EMPTY*

Pressure build up

*Petrol Tank FULL*

**Hobson Telegauge - Operation**

unit. Slacken off the union nuts and gently break the seal of the tapered fitting on the end of the pipes. Use a drill bit rotated by the fingers to clear the pipes of sediment.

For brass union - use No 54 or 1.4mm drill.

For copper pipe - use 3/64" or 1.2mm drill. For cleaning the pipe between the brass reservoir and the glass tube, use a piece of cheese cutting springy wire or similar.

### Pipe identification

*Main Air Line* - thickest pipe and largest union.

*Pressure Balance Line* - medium sized pipe.

*Air Pump Line* - thinnest pipe, smallest union.

When both tank and instrument ends of the fine bored delivery tube are clear, plug one end and suck on the other. The vacuum so created should hold the tongue stuck firmly to the tube end. Then clear the tube of all moisture, by blowing compressed air through it. It is essential that the line is clear and dry - don't blow through it as your breath will condense and leave a film of moisture. Do not use an air compressor as the pipe-work is too delicate. Use a bicycle

bottom of the air chamber, resulting in a reading being obtained in the sight glass.

A non working gauge, is usually due to evaporation of either the fluid, or a pressure leak in the line between gauge and the petrol tank unit. In practice, any leaks are usually at the connections. The petrol tank end is self explanatory; check tightness of the coupling and it is recommended to use a thin film of sealant on the threads. Be careful with the dash instrument connections. If possible remove the instrument from the dash, before attempting to remove the pipe connections. Be careful not to kink the tubes when removing and under no circumstances put any stress on the fragile die cast body of the instrument.

If the gauge has been empty of fluid for some time, it is likely that the pipes are blocked with sediment at the sight gauge pump and cut off the connector from the delivery hose. Push the hose over the air line at the gauge end and give at least 50 full strokes; some petrol may be expelled from the Pressure Balance line.

Reconnect the line to the petrol tank and pump more air through, while someone listens for bubbling at the filler neck of the petrol tank. This will prove that the line is clear. If the line is kinked or broken, it is possible to repair, by bridging a broken pipe with a piece of suitably sized polythene tube. Refill the instrument (with the air-line disconnected at the instrument end) by pouring Tetrabromoethane with an eye dropper into the 'U' tube, until the liquid level reads 'zero'. The air-line is then re-connected to the gauge and the accuracy checked against a known quantity of fuel in the petrol tank. The gauge reading is adjusted by adding or removing one or more calibrating wires. These wires which were put in when the instrument was first commissioned, are unlikely to need changing. Unless, of course, you are trying to achieve a 100% accurate read out.

If the gauge is dry and has been for a long time, it is probable that the sediment visible in the sight gauge, has also blocked the fine copper pipe leading to the sight gauge. This pipe is too fine to clean and will need replacing. Vintage Restorations in Tunbridge Wells can do this for you and at the same time will ultrasonically clean the sight glass. The other problem you might encounter, is degradation of the instrument

facia. The lettering and scale was originally screen printed on formed aluminium and would be very expensive to replicate on a one-off basis. I reproduced the scale on my Apple iMac, printed it on thin card, sealed it with a clear polyurethane spray and then stuck it over the aluminium. The reassembled gauge looks pleasingly authentic.

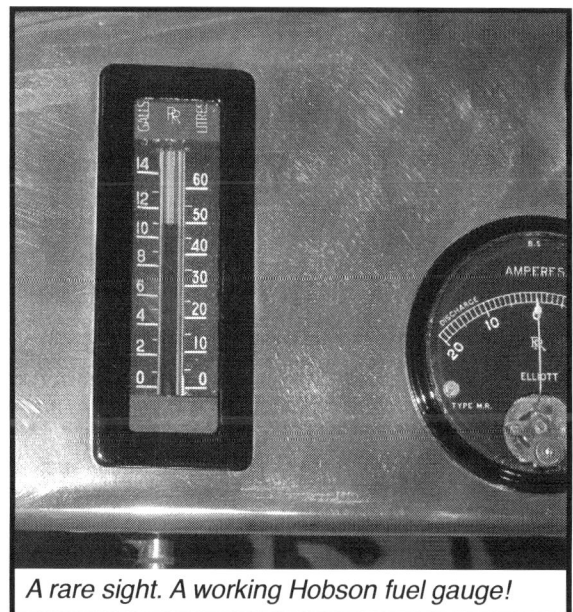

*A rare sight. A working Hobson fuel gauge!*

## Theory

The ignition system fitted to the Rolls-Royce 20HP and early 20/25 is a conventional coil system. When the ignition is first switched on, battery voltage (circa 12v) is applied across the ignition coil. After a few seconds, a ballast resistor warms up and this reduces the voltage to 4 volts. The coil is designed to operate at this lower voltage; so when starting from cold, the coil is 'over-voltaged' and therefore produces a bigger, hotter spark to facilitate a quick engine start. Once this is appreciated, you will understand why the correct way to start the engine, is to operate the starter and only *then*, to switch on the ignition. So many owners, switch on the ignition and then fiddle about with the starting and mixture controls. By the time they operate the starter, all the benefits of an over-voltaged hot spark have been lost.

When the engine is running, or being turned over on the starting handle, the voltage feeding the coil is switched on and off by the action of the contact breaker within the distributor. The coil can be thought of as a transformer. Inside is a winding of relatively thick wire called the 'primary'. It is this winding, or 'coil', that the current flows through. Many more turns of thinner wire are wound on top of the primary and

Original Rolls-Royce coil (bottom right). Note: coil is not in use, HT lead is in its earthed holder. Magneto HT cable is plugged into the distributor. When running with the magneto, it is essential that Fuse No 3 in the fuse box (top left) is removed. There is a holder for the removed fuse, inside the cover of the fuse box. The ignition ballast resistor, is to the left of the fusebox.

this coil is called the secondary. Both coils are wound on a common soft iron core. When the current that is flowing in the primary of the coil is interrupted by the action of the contact breaker, the collapsing magnetic field induces a voltage across the coil, of opposite polarity to the supply. The coil in fact, becomes a source of supply instead of a load. The voltage induced in the secondary coil, is dependent on the relationship of the coil turns in the primary and secondary. In broad terms, if the coil turns are in the ratio 1:2, (this is hypothetical and given purely as an aid to understanding) the primary coil might have 50 turns and the secondary coil 100 turns. In this example, when 4 volts at 2 amps is flowing in the primary and is interrupted (switched off), then 8 volts at 1 amp is induced in the secondary. From this example it will be seen, that the bigger the turns ratio in the two coils, then the bigger the voltage induced in the

secondary. It is very high voltage that produces a spark, of sufficient intensity to fire the fuel mixture. So the fact that the coil secondary also steps down the current, is of little consequence.

In practice, the turns ratio in the coil is of the order 1:60. So 12 volts (cold) applied to the primary will produce circa 720 volts in the secondary winding. However, this is still not enough voltage to generate a fat enough spark at the spark plug; more on this later. A variable to be kept in mind when considering the voltage required to produce a fat spark, is atmospheric pressure. As pressure is increased, so a higher voltage is required to create a spark. This is why

*Coil winding 'station' at Magneto Repairs in Stroud, Gloucestershire. This firm rewinds both Rolls-Royce coils and magnetos.*

competition engines that run high compression ratios, need a 'sports' coil to provide the higher voltage that is required at the plug. Because of this phenomenon, it is also the case that it may be possible to see a spark at the plug, that would be absent when the plug is subject to compression pressure, ie: when the engine is actually running. Similarly, a weak spark may be sufficient to run the engine at low throttle openings, but will cause misfiring at higher throttle openings, as high throttle openings produce more compression within the combustion chamber. The engine needs a minimum voltage at the plug of circa 15,000 volts to spark at all; though in fact the system is designed to deliver considerably more than this.

The stepped up HT voltage from the coil is passed down the HT lead to the distributor, where the revolving rotor disk delivers it to the correct spark plug. 'Correct' in this sense, means the plug that needs to be sparked in the cylinder, where atomised fuel is ready to be burnt. The Rolls-Royce engine firing order is 1 - 4 - 2 - 6 - 3 - 5 (with 1 being the cylinder nearest the radiator) and this is the order that the HT leads from the distributor go to the plugs. But how do we get from the circa 720 volts generated in the secondary of the coil, to the very high voltage that is necessary to produce a good spark under compression? The answer; is from the resonant circuit formed by the coil and condenser.

The easiest way to understand a resonant circuit, is to consider the case of a steel ruler held flat at one end by the palm of a hand on a desk and the other end 'twanged'. The ruler rapidly oscillates up and down and a note is heard; this is a simple resonant circuit. The circuit consisting of the coil and condenser, is the electrical equivalent of the vibrating steel ruler. The 'twanging' occurs when the current flowing in the coil circuit is interrupted. This interruption has to be as instantaneous as possible and the action of the contact breaker opening on its own, is not 'fast' enough. This is where the condenser comes in. The condenser is situated within, or on the distributor and has two main functions. To 'boost' the coil

*Rotor disk. The brass finger is showing signs of spark erosion at the end. New fingers are available. The other screw/washer does not hold anything. It is simply a counterweight to the finger, to ensure a stable spin. Attention to detail!*

voltage and in so doing, to absorb the energy at the contact breaker as they open (switching) and so prevent a secondary spark across the contact points.

The coil spends most of its time being charged with energy (current) from the circuit provided by the closed contact points. This creates a steadily rising magnetic flux acting on the soft iron core within the coil. This current goes from zero to several amps and takes at least three milliseconds to charge the coil to this level. The relatively long build up time is important in eventually producing a fat spark at the plug and which is why correct 'dwell' is important. The moment the points open, the current that was flowing in the coil, is instantly diverted to the condenser. The stored energy within the coil's magnetic field is transferred into the electric field in the condenser causing the coil current to rapidly decay to zero, while at the same time, condenser voltage rises to a maximum, as a function of the

*Bluebell on the 20HP Rally to The Loire June 2013.*

resonant circuit. If the condenser is faulty, then the opening of the points, causes the charged coil to try to bridge the points gap and results in a spark at the points instead of the plugs and which will quickly erode the points. When the condenser is present and sound, then when the points open, to all intents and purposes the positive supply and the chassis earth are connected with only the internal battery resistance between them, creating a coil and condenser in a complete circuit that oscillates for a few cycles until circuit losses damp it out. By the time the points close, the condenser has been completely discharged and there is no energy left to cause unwanted sparking at the points.

Coil current is now at zero and the condenser has been charged to many hundreds of volts, because of the coil inductance. Then the process reverses. The condenser dumps its energy back into the coil until the condenser voltage is once more at zero. This effect of a resonant circuit is called 'ringing'. Each successive

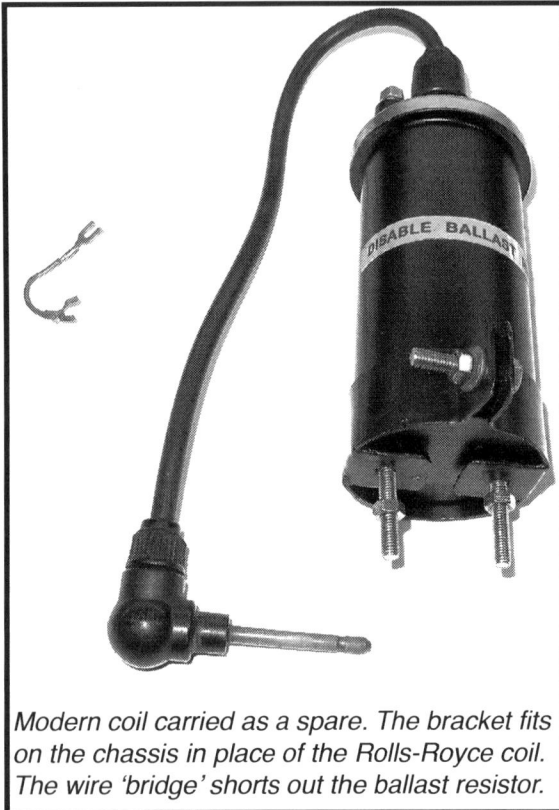

Modern coil carried as a spare. The bracket fits on the chassis in place of the Rolls-Royce coil. The wire 'bridge' shorts out the ballast resistor.

'ring' is smaller than the one before and dies away after a few cycles, depending on the losses in the circuit etc., in the same way that the vibrating ruler rapidly comes to rest. The effect of this 'ringing', is that a short lived burst of AC current at circa 500 volts is induced in the primary of the coil and is transformed into circa 30,000 volts in the secondary of the coil. The 'spark' that then occurs at the sparking plug is not in fact a single spark, but a short lived burst of sparks at high energy and frequency.

It is interesting to note, that 'ringing' also occurs on the HT cable side. When the spark plug fires, it also discharges the capacitance of the plug leads and although this value is small (ringing frequencies of megahertz instead of the kilohertz found in the coil/condenser circuit), repeated sparking occurs. It is this secondary sparking at the plug that generates radio interference.

### Ballast Ignition fitted to Rolls-Royce

There are two main benefits to be had from the ballast system. The first, is that in a normal coil ignition system, the spark tends to get weaker as engine revs rise. This is because when the points close, current from the battery flows through the coil, and because of the property of the coil termed inductance, current rise is slowed. In fact, if the current rise is plotted on a graph, it displays a curve and takes a relatively long time to reach peak value. In practice, peak value is never actually reached before the points once again open and switches the current off. Because the actual time that the points are closed (dwell) becomes less and less as revs increase, the energy in the coil and therefore the spark intensity also becomes less as engine revs increase. So, how does a resistor feeding the coil improve this state of affairs? When the points close, and the coil begins to charge, the voltage

70

dropped across the resistor (and which is proportional to the current flowing through it) increases, so the actual voltage present at the coil reduces, in turn reducing the current. The net effect of this, is to flatten the top of the current curve, so that differences in dwell time make less difference to current.

The other benefit of the ballast system, is that advantage can be taken of the fact, that the Rolls-Royce coil is designed to operate at circa 4 volts. Therefore if battery voltage at 12 volts could be made available to the coil when the starter is cranking the engine, then this short period of extra voltage will produce a fatter spark. The ballast resistor is wound from a high temperature coefficient wire. This means that when cold, it has a lower resistance, therefore supplying close to battery voltage to the coil and producing a higher intensity spark. When hot, the resistance rises and reduces the voltage fed to the coil, to circa 4 volts.

### Servicing the ignition system

Although the underlying physics of generating a spark is quite complex, the system in practice is relatively simple. The most important aspect, is maintaining the correct setting of the contact breaker, as this determines 'dwell'. All HT leads should be clean, in good condition and with properly secured terminals; sporadic misfiring is often a fault in one or more HT leads. This can sometimes be seen, by blipping the throttle while watching the HT leads in a darkened garage. Any unwanted electrical tracking will be seen and often heard. If in doubt, replace the HT leads. Copper cored braided HT cable, PVC insulated, can be bought mail order from: Auto Electric.

### Condenser

The ignition condenser value is 0.47mfd. A modern tubular version to fit on

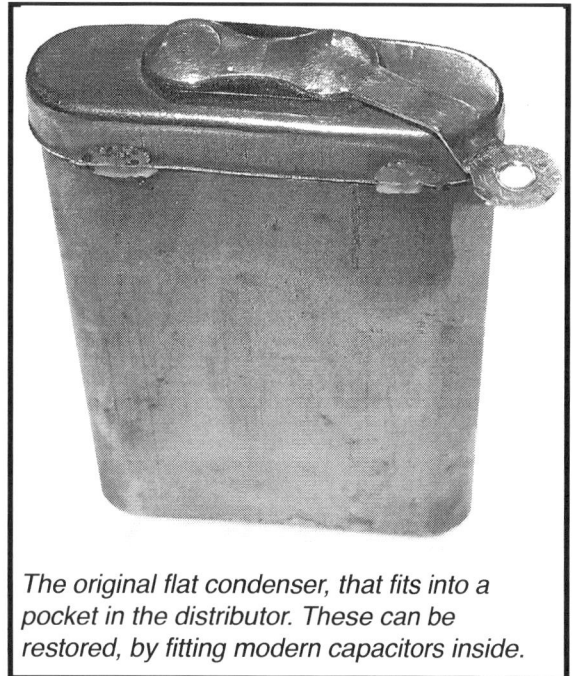

*The original flat condenser, that fits into a pocket in the distributor. These can be restored, by fitting modern capacitors inside.*

the outside of the distributor is also available from Auto Electric Supplies. The correct flat condenser to fit inside the distributor, is available from Ristes. A lower value modern condenser may be used in an emergency situation. Engine performance will suffer, but it will get you home. The original flat condenser can also be restored, by fitting modern capacitors inside. A typical re-build should show: 0.296 mfd. with zero leakage at 500 volts DC. The modern capacitors used, are rated at 110 degrees C and 275 volts AC. Mine was rebuilt by David Else, who also supplied the condenser data above.

Badly corroded contact breaker points on a car with a weak or non existent spark, is an indication that the condenser is breaking down. A specialised testing rig is required to correctly test a condenser. A quick test is to remove the condenser and apply battery voltage between the terminal and the case. Disconnect the battery and put test leads from a multimeter across the condenser. The needle on the meter should show a strong 'flick as the stored energy is discharged by the meter. If in doubt, it is wise to test by substitution. Many a 'mystery' ignition fault has been cured by the fitting of a new condenser.

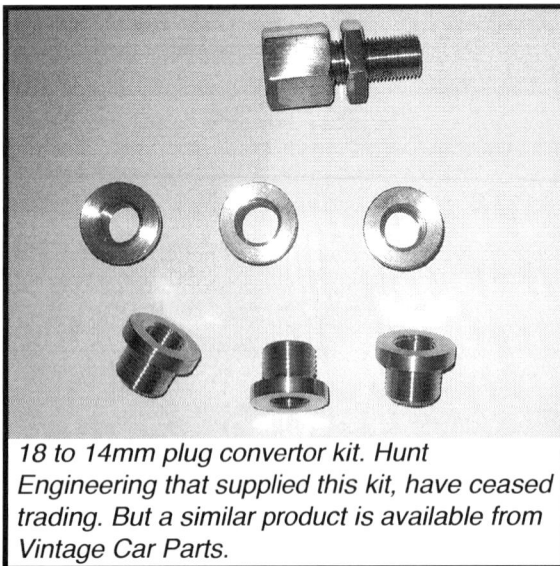

18 to 14mm plug convertor kit. Hunt Engineering that supplied this kit, have ceased trading. But a similar product is available from Vintage Car Parts.

## Coil

This component rarely gives trouble. To test: turn the engine over until the contact points are seen to be closed. Remove the HT bayonet connection at the distributor (leading from the coil). Turn on the ignition. Position the naked bayonet 3-4 mm from earth - the water pump is a handy place. Use a small screwdriver to 'flick the points open. If the coil is OK, a spark will be seen. If the spark looks weak, suspect a faulty condenser.

If the coil fails, the answer is to get it rewound at Magneto Repairs. This firm re-winds and impregnates the windings with wax. *Do not use any firm that uses varnish* 'as Henry Royce did'. If the coil ever fails again, trying to remove the old varnish usually destroys the coil former.

Finally, if you are a belt&braces man, then it's not a bad idea to carry a cheap modern coil, for use in case both the coil and the magneto ignition fail. You can fit any modern 12v negative earth coil; make up a bracket as shown in the picture, to take the place of the Rolls-Royce coil. A nicety is to fit a spare bayonet connector, so that it is immediately ready for use. Finally stick a reminder on the coil to disable the ballast resistor.

I carry a 'bridge', shown in the picture, that fits under the thumb screws on the bulkhead mounted ballast resistor. Note that when using a non-ballast 12v coil, the original Rolls-Royce condenser will be the wrong value. But it will get you home. The correct condenser to use, should be in the range .18-.25 mfd.

### Rotor Arm

The physical distance between the screw holding the rotor disk 'finger' and the distributor shaft, which is at earth potential, is quite small. Over time, the bakelite from which the rotor is made, ages. HT eventually leaks through the rotor disk to earth. Initially, this will present as a mysterious misfire and/or lack of power. Eventually, there will be no spark at all. So if your engine still has the original distributor cap and rotor disk in place, it is likely that the insulation is degrading, leading to rough running. This is sometimes weather related. Rolls-Royce manufactured their own bakelite components on the early cars, using early material, that is now past its sell-by-date. Many an elusive misfire has eventually proved to be a 'leaky' cap or rotor disk.

The rotor disk is also often the source of another mystery misfire, caused by an overlarge clearance between the 'finger' and the HT terminals inside the distributor cap. The clearance between the rotor and the terminals should be circa 2 thou. If outside this range, fit a new finger. A temporary fix, is to remove and peen the brass finger, then file down the contact end to achieve the clearance. Fiennes can supply a new distributor cap and a complete new rotor disk. Neither items are cheap, but they are manufactured from modern materials and will prove more reliable. The brass 'finger'

on the rotor disk, is also available separately.

### Sparking Plugs

The 20HP and 20/25 were originally fitted with a half inch reach 18mm spark plug, such as a Champion D16. This size of plug has been obsolete for a long time now, though there is some new 'old stock' Champion No 7 that can be found. These old stock plugs were made in Canada for the Military, during WWII and are old technology. A better solution, in my opinion, is to fit a plug convertor; available from Vintage Car Parts. This will allow you to use a modern 14mm three quarter inch reach plug. The modern plug most suited to

*The flywheel revealed, after the floorboards and the clutch access cover are removed. You need to rotate the engine from here, to align the pointer with the BLI mark (early cars) or the MAI mark (later cars). Room for confusion, as the magneto timing mark is also displayed, as well as valve timing marks. Use a torch, identify the correct mark & dab it with a spot of white paint.*

73

the early cars, is the NGK BP5ES. Not only a better plug than the old half inch plugs, but longer lasting. If a slightly cooler plug suits your engine better, then the NGK BP6ES is called for. These plugs are cheap enough to afford to experiment.

### *Timing - coil ignition*

1.      Follow the plug lead from No 1 spark plug (nearest to the radiator) back to the distributor. Note the position of the lead in the distributor cap. Lift off the cap and put a dob of white paint on the distributor rim, to mark the position of the rotor, where it fires to No 1.

2.      Rotate the engine, on the handle, with the plugs removed, until the rotor arm is approaching the No 1 cylinder position. Then rotate further, by levering with a long screwdriver on the bolts holding the clutch, inside the flywheel inspection chamber. Alternatively, jack up the rear and rotate a road wheel (in gear) to turn the engine. When you turn the engine on the handle, it has the effect of 'winding up' the slipper drive and will result in a false timing position. So you have to turn the engine, by applying the force to the back of the crankshaft and not the front.
       Turn the engine, as described, to align the white pointer at the top, with the BLI mark on the flywheel. Note that BLI stands for 'Battery (ie:coil) Late Ignition. Later

cars are stamped MAI (Maximum Advance Ignition).

3.      Move the steering wheel ignition control DOWN to fully Late for a BLI mark. Move the control UP to full early, for a MAI mark. Remove the rotor arm. Note its position on the cam it fits on. Loosen the screw holding the cam on its tapered shaft. Carefully lever upwards on the key on the cam, to release the cam. But keeping the same general position of the cam, so that when the rotor arm is put back on, it is still pointing towards No 1 cylinder. If you remove the screw completely, there is a danger of the cam 'pinging' off and getting lost, when it is levered off.

4.      Turn the cam anti-clockwise until the points just open. The most accurate way to do this, is to connect a test lamp between the coil and the low tension connection to the contact breaker. The light will go out as the points open. Then, without altering the position of the cam, press it down to lock on the conical shaft and tighten the holding screw.

The magneto system is Henry Royce at his finest. The ignition system was the achilles heel of all early motor cars. Henry's solution, was to fit both coil *and* magneto ignition. The brilliance, was not so much the foresight to fit a secondary ignition system, but to make both systems quick and easy to change over. Today, most early Rolls-Royce cars, have had both the coil and condenser rewound/replaced using modern materials and consequently coil ignition is now very reliable. Which is why under so many highly polished bonnets, you will find either a magneto that no longer works, or is not timed and so cannot work. Which is a pity, as the magneto is not difficult to reinstate and one day when your car "fails to proceed", may well get

you home.

## Checking that it is working

Engage the magneto. Undo the thumb screw on top of the magneto and release the HT cable. Refit the screw with a short piece of wire underneath. Position the end of the wire a couple of mm from any nearby metal (earth). Get an assistant to spin the engine over, either on the starter or the handle, while you watch the spark gap at the end of

A - oiler showing direction of magneto rotation. **B** - HT output to distributor. **C** - earth terminal on end cap. **D** Remove this nut and then turn control on steering wheel to full 'early', when bolt can be withdrawn to disconnect the advance/retard lever. **E** - Use correct C spanner to undo drive coupling; collect spring from inside and large flat sealing washer. **F** - One of two bolts that holds the magneto to the engine; the other bolt is diagonally opposite and is the longer of the two. **G** - Control rod to early/late on steering wheel.

the wire. Do not switch on the ignition. If you see a spark, the magneto is almost certainly OK. Reconnect the HT cable and repeat above, but this time with the HT bayonet connector, after you have pulled it out of its housing in the base of the distributor. If you again see a spark, then the HT cable that runs around the back of the engine, is also OK. If there is no spark in either case, then the magneto is at fault.

### Adjusting & bench test

Remove the magneto to the bench. There are two bolts underneath; note they are different sizes. The coupling is undone using a 'C' spanner that was part of the original tool kit. Lastly, there is a small nut & bolt holding the control rod at the top that needs to be removed completely. Operate the early/late control on the steering wheel, to move the control rod to get access to this fastening. The magneto will then lift off the

Mark the pointer and the MLI mark inside the clutch cover, with a paint dob. Use separate colours for the coil mark & the magneto mark.

platform it sits on, complete with the operating shaft. Then remove the shaft.

Clean and adjust the points to .014 - .018 thou and re-test. Connect a piece of wire to the thumb screw and position it a few mm from the metal body. Flick the magneto over by twirling the input connection with your fingers. If there is still no spark, the probability is that the magneto needs rewinding. Apart from the points, the only adjustment is by way of two screws which govern the initial position of the advance/retard operating arm. If the magneto has previously been working and this arm has not been removed - then leave it alone as it will already be in the correct position! Send the magneto to Magneto Repairs in Stroud, who have the necessary expertise to rebuild.

### Timing the magneto

*Note: the following does not apply to the very early magneto, which features separate leads to the individual spark plugs.*

**1.** Remove the plugs (to make turning the engine over easier) and the distributor cap. Rotate the engine until the rotor arm is pointing at No 1 cylinder plug lead; this is the cylinder nearest the radiator. Mark this position with a dab of white paint on the inside rim of the distributor. Rotate the engine again, until the rotor arm is approaching the paint dab. With the front floorboards and the inspection panel on the flywheel housing removed, rotate the engine further, by levering with a long screwdriver on the bolt heads holding the clutch. The

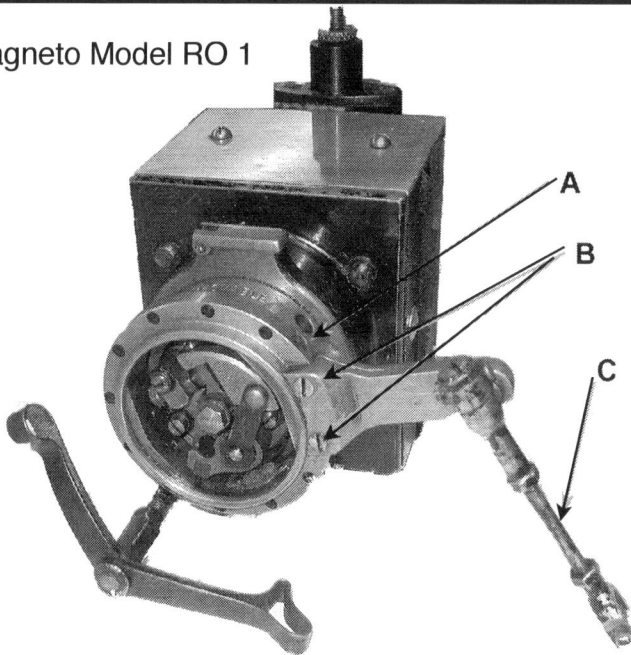

Watford Magneto Model RO 1

*'A' is not an adjuster and should be ignored. 'B' is to get approximate timing and moves the operating lever in 12 fixed increments. Rotate the operating rod 'C' after slackening the two lock nuts, to achieve full early/late*

2. Move the early/late control on the steering wheel to fully 'late'. Place the magneto back in position on the engine and turn the operating arm of the early/late control, so that it registers with the position of the engine has to be rotated independently of the starting handle, to avoid the slipper vibration damper inside the engine 'winding up' and thereby delivering a false timing position. The tiny pointer visible at the top of the inspection chamber, needs to align with the TDC (Top Dead Centre) or MLI (Magneto Late Ignition) marks on the flywheel. Note that this mark is stamped vertically and so your head needs to be at 90° to read it; it makes the job easier, if you mark the pointer with white paint as well. Do not be confused by the mark BLI which is further around the flywheel and is the mark for timing the Battery (coil) ignition system.

(disconnected) operating rod.

3. With the magneto on the bench and with the shaft in position, turn the magneto drive until the points are fully open. Insert the thinnest feeler gauge (or a cigarette paper) and turn again until the gauge or paper is gripped by the closed points. Then turn the magneto drive again, in the correct direction of rotation (anti-clock looking at the points and marked by an arrow on top of the oiler door) until the gauge is just released. Mark this position with a dab of paint on the rim of the magneto, aligned with the points. Use a strip of thin rubber or cardboard to act as a wedge and to jam the points in this just opening position.

**4.** Engage the magneto drive shaft with the coupling on the dynamo. Ensure that the drive shaft is locked in the engaged position; you will feel the shaft engage as you rotate it by hand, pressing it in towards the dynamo. Offer up the magneto to the 'locked' drive shaft on the engine, not forgetting the spring you collected when the magneto was removed. Just before the drive cog engages, check that the position of the just opening points is still correct - as shown by the paint dab. Note that the points position does not have to be 100% accurate - as the steering wheel control lever will allow for a little discrepancy. Each 'click' on the steering wheel control, represents one degree of ignition advance or retard - don't you just love Rolls-Royce precision!

**5.** Engage the drive cog and slip the two magneto holding bolts home and then check the position of the points again. Assuming they have not moved out of position by more than a degree or so - that's it. Tighten the two bolts and the drive coupling, reconnect the early/late operating rod, fit the end cap & cable, start the engine and test.

According to the Chassis Card data, using the magneto as opposed to the normal coil ignition, results in a small power loss. I can't say this is noticeable during real world motoring with GXO71; if anything the engine runs sweeter than ever on the magneto!

*Caution:* Whenever starting the engine (especially on the handle) with the magneto engaged, always move the steering wheel control, at least half way to fully late, to avoid an engine 'kick back'. It is also essential to *remove No 3 ignition fuse from the fuse box*. This disables the coil ignition when running the engine on the magneto. If the battery is disconnected, then this fuse can be left in position. *But on no account set the ignition switch to 'charge' with the engine running and the battery disconnected*. The Handbook advises playing safe and disconnecting the field connection at the dynamo.

Also be aware, that when running on the magneto, do not fire up the engine with a plug lead disconnected. This can lead to high damaging voltages within the magneto.

Finally; never engage the magneto drive with the HT lead from the magneto plugged into the base (earth) of the distributor. With coil ignition engaged and the engine running, this will short the magneto and melt the windings.

Trying to adjust or replace the points with the distributor on the engine is a bit fiddly. Far easier, is to remove the whole distributor to the bench. But before doing so, mark the position of the cam, to avoid having to re-time the engine. Start by marking with a dab of white paint on the rim of the distributor, where the finger of the rotor disk is pointing. Remove the rotor disk. Clean the cam and the inside of the distributor. Then take a piece of clear tape and rule a straight line on the sticky side, using a marker pen. Then stick the tape over the junction with the cam and the distributor

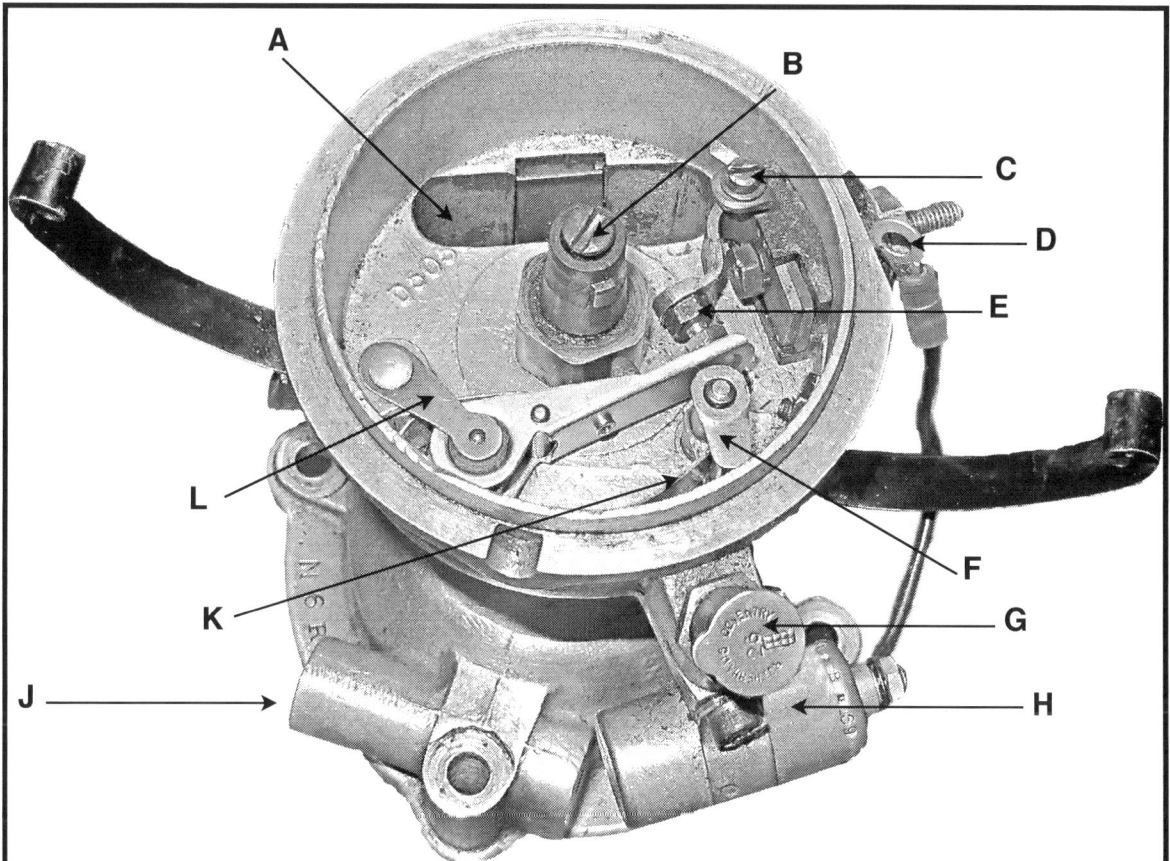

### Modern condenser modification

A- This is where the original condenser was fitted and which later became obsolete. The cylindrical condenser at (H) is an approved Rolls-Royce modification. B - Slacken this screw and lever up to remove the cam. C - This is where the original condenser connected to the points. D - Low tension & new condenser connection. E - Slacken this locking screw to allow points adjustment (17 to 21 thou). F - Lift off pivot arm that locates the leaf spring anchored underneath (K). G - Flip top oiler to advance & retard mechanism. H - Modern tubular condenser. J - Parking hole for HT Magneto plug. K - Soft metal strip that runs parallel with leaf spring and is attached to the rocking arm of the points. It earths here via a plate held by a screw on the side of distributor.

and then with a sharp blade, cut along the base of the cam, so that it is free. Slacken the screw at the top holding the cam, *but do not remove it.* The cam can then be pried up and free of the taper it sits on. Then remove the screw, which has stopped the cam flying off when it was pried up and lift the cam off. I bought two cheap electrical screwdrivers and after heating to a cherry red, bent the tips to 90 degrees. These slip underneath the cam and quickly and easily pops it off the taper.

Undo the four bolts that hold the distributor to the engine casing, the advance/retard control rod and the cable from the coil. The distributor can then be gently pried up and off. When it is off, you will see that there is a slot in the driving cog; this ensures the distributor can only be put back in one position. If you have purchased new contact breakers, be aware that there are two types that have been manufactured. One type is as per the original design. But there is another, that has a deeper fixed point and needs the securing screw moved to the other side of the mounting bracket. See the detail in the relevant picture caption.

If you have a mechanical problem with the distributor, it is usually a worn top bearing, leading to a sloppy cam and erratic timing. Unless you have the specialised tools to undertake a mechanical overhaul of the distributor, you would be well advised to send the distributor to an engineering workshop. I had my distributor completely rebuilt by Fiennes and which cost circa £320 exc. vat.

*Mark a line on the sticky side of a piece of tape, to enable the cam to be replaced in exactly the same position.*

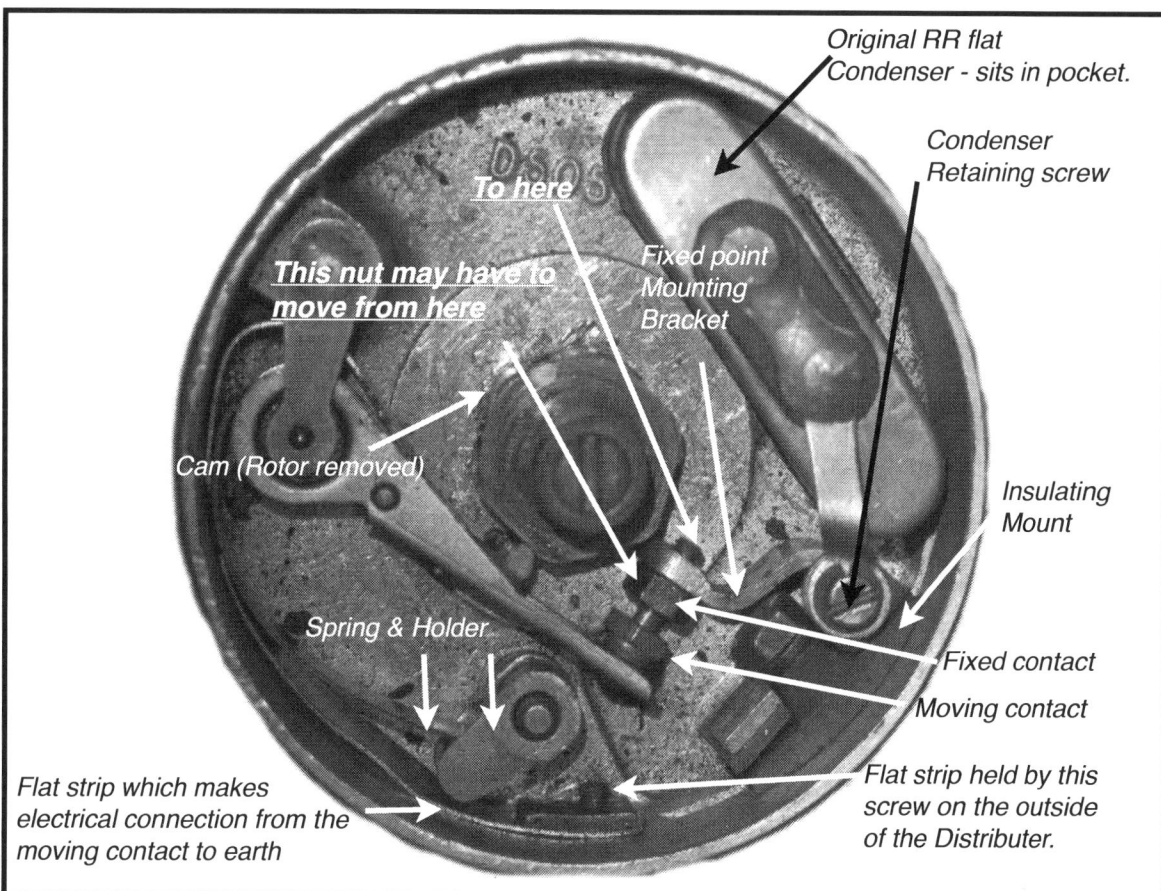

**Original RR flat Condenser - sits in pocket.**

**Condenser Retaining screw**

**To here**

**This nut may have to move from here**

**Fixed point Mounting Bracket**

**Cam (Rotor removed)**

**Insulating Mount**

**Spring & Holder**

**Fixed contact**

**Moving contact**

**Flat strip which makes electrical connection from the moving contact to earth**

**Flat strip held by this screw on the outside of the Distributer.**

**To replace the points.** *The easiest way is to remove the distributor and mount it in a vice with soft jaws. Before doing so, mark the position of the rotor; that will help to locate the drive shaft when replacing. Remove the 4 nuts that holds the distributor and pedestal and the collar/split pin that secures the end of the advance/retard control rod. Remove the distributor. Accurately mark the position of the cam. Then remove the cam by slackening off the screw that secures it and then levering the cam upwards. By leaving the screw in place, you remove the risk of the cam flying off when levering. Then remove the screw and the cam. Remove the condenser which is an interference fit in the distributor body. Remove the fixed contact point being careful not to break the insulating mount the securing nut sits on, as it is fragile. Remove the moving point, by swinging the securing arm off the pivot and then levering the contact arm and spring holder off their respective posts. Note that the spring is separate and is re-used. Fit the new points in reverse order to dismantling. Note that the fixed point has been supplied in the past with a 'thin' actual contact and is held and secured by the nut as shown in the picture. If the new fixed point as supplied has a point depth deeper (circa 4mm) - then the securing nut needs to go on the other side of the mounting bracket as indicated in the picture.*

## Starter Motor Switch

If you find there's a reluctance at times for the starter motor to operate, it's possibly a faulty starter motor switch. Lift the nearside bonnet panel and you will find this mechanical switch on the bulkhead, just above the starter motor. On top of the switch is a small bolt; remove and fill with oil, using a pump action oil can. This is often overlooked during the annual service and will result in burnt contacts. A lack of oil will also cause erosion of the internal copper contact fingers. The copper cage and the fingers that make contact and complete the circuit can be seen at the top of the picture. Clean everything up, re-tension the copper fingers and fill the unit with oil.

*Fred Snoxall with his son Roy, photographed in Boxmoor sometime in the 1930s.*

*Dismantled starter motor switch. The rod on the bottom piece, pushes the circular piece, into the copper cage, visible in the top component. This completes the circuit.*

The battery is carried in a box, fitted to the chassis. Early cars have wooden boxes, while later cars have metal ones. The wooden boxes seem to have stood up to the test of time, rather better than the metal ones.

Battery technology has improved considerably, from when the cars were new, when a 50 amp/hr battery was fitted. A new battery today, with the same physical dimensions to fit into the battery box, will be rated at 96 amp/hr and give greatly increased performance. All models feature cotton covered braided wiring and by now, all cars will either have been re-wired or will need rewiring. Modern plastic insulated cable, covered with the correct cotton braiding is available in a variety of colours, from Auto Electric Supplies. The original Ross-Courtney cable terminals are available from

Fiennes. DIY re-wiring is within the scope of a methodically minded amateur. I chose to have my 25/30 re-wired at a reasonable fixed price, by Phil Cordery a professional Rolls-Royce auto electrician.

It would be wise to fit a Battery Master Switch; especially if the wiring on your car, is showing signs of degradation. It is also useful, to be able to leave the car with all electrics disabled. Especially if you are leaving the car parked in a public place. There are two options; a physical switch and one operated wirelessly via a key fob. The latter is made in the US under the name

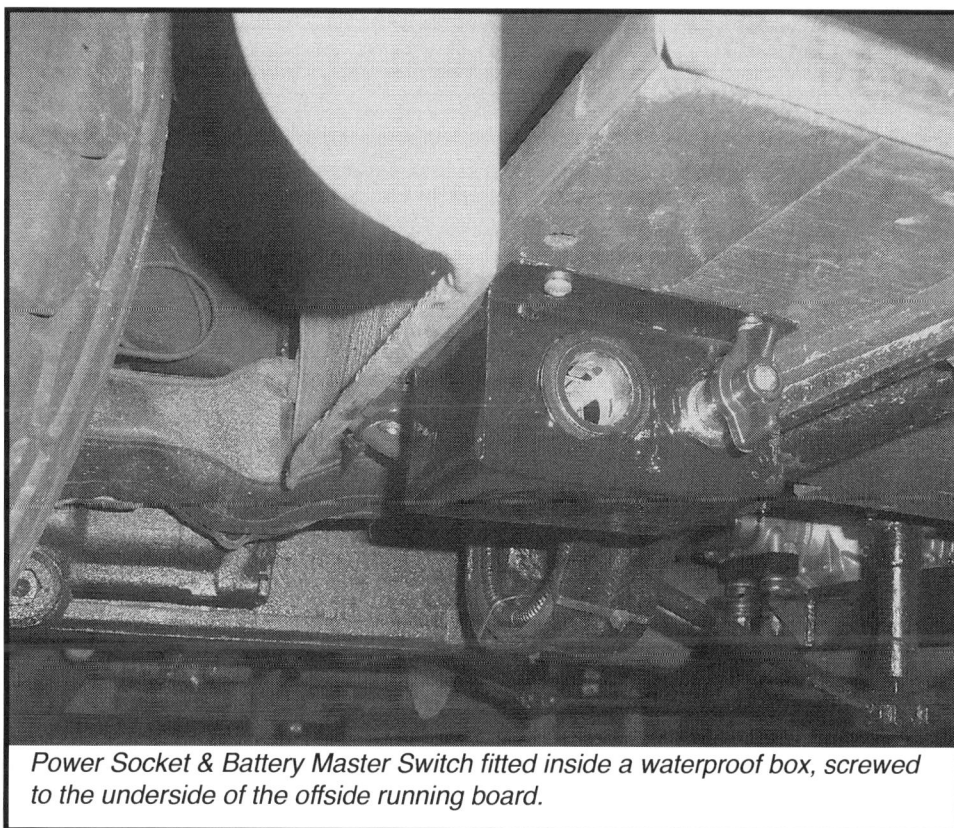

*Power Socket & Battery Master Switch fitted inside a waterproof box, screwed to the underside of the offside running board.*

*An extra 12 way auxiliary distribution/fuse box fitted under the dash, of a 25/30. This is to power an Overdrive, SatNav socket, Flashing Indicators, additional Power Socket etc.*

the rear wing. I took the opportunity to fit a power socket here at the same time, wired directly to the battery, via an in-line cartridge fuse. This means that the car can be left for long periods with the battery isolated, yet at the same time left connected to a trickle charger. Without all the bother, of removing the floorboards, to get access to the battery. Both switch and power socket are contained in a metal box, fabricated from 18 gauge sheet steel. Some owners, who like the comfort of being able to kill the battery from the driver's seat, fit the switch in the floor, near the gear lever.

'Battery-Brain'. The physical switch is widely available, including from Auto Electric. Ideally, it needs to be fitted as close to the battery as possible. On cars with running boards, a handy place is underneath the offside board, just in front of

*Wooden battery box, early 20/25. The armoured cables running between the body and chassis rail, is a modification. Used to connect to the Battery Master Switch and the (fused) Trickle Charger Power Socket.*

The early small Rolls-Royce chassis is fitted with a three brush dynamo; easy to identify, as it has three cables connected to the dynamo. This early dynamo is nothing like as efficient as the charging system on a modern car. Unlike the modern car, as the speed of the RR engine rises, the charge from the dynamo does not. In fact, with a battery that needs charging, increased speed actually decreases the output from the dynamo, due to distortion in the field coil's magnetic flux.

The third brush in the dynamo is physically smaller than the other two. It is used to pick up current from the revolving commutator, to power the field coil, which magnetises the soft iron poles. The third brush is adjustable, the physical position determining the output to the field coil. This is adjusted when the dynamo is initially set up and should not be altered, unless you know what you are doing. To increase the charge rate, move the field brush in its holder towards the direction of rotation of the dynamo, to decrease, move the other way.

The major shortcomings of the third brush system, is that it is possible to either overcharge the battery on a long run, or to flatten the battery with a lot of short stop and start runs. Broad brush advice, is to switch on the charge for about five minutes in any given hour, on a daylight run. However, if you have an Overdrive and/or flashing indicators fitted, the former draws

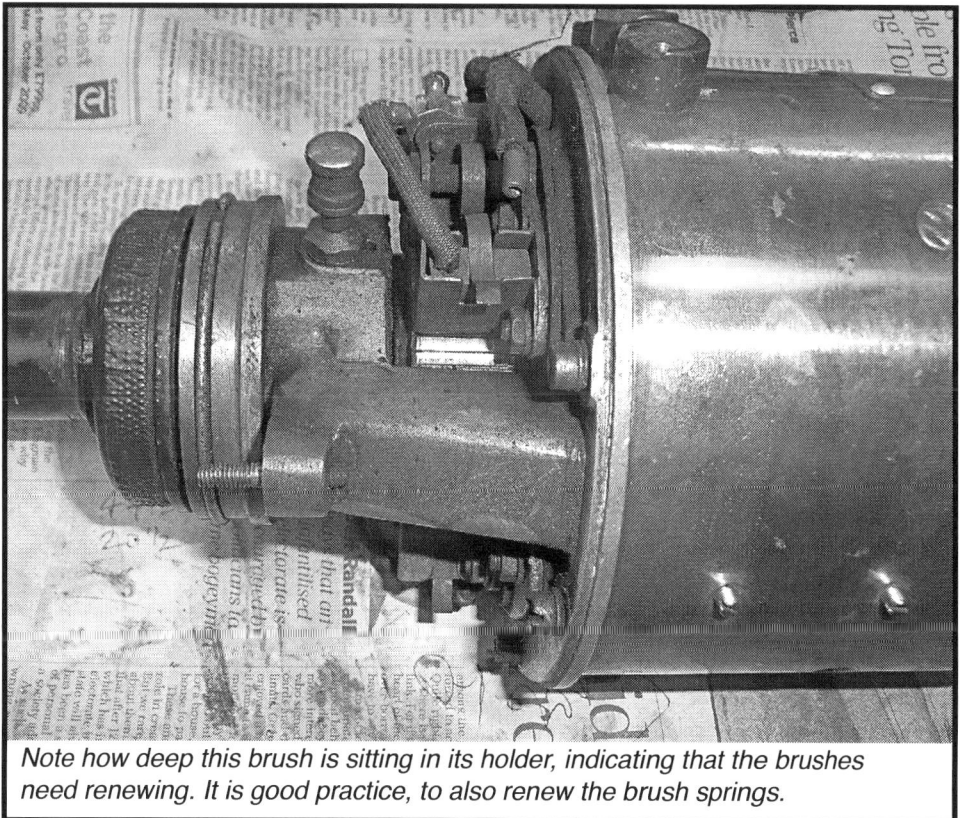

*Note how deep this brush is sitting in its holder, indicating that the brushes need renewing. It is good practice, to also renew the brush springs.*

about 2 amps when in use, in which case charge for ten minutes every hour. At night with lights on, leave the dynamo on charge permanently. Also be aware, that a three-brush dynamo takes a very long time to recharge a completely exhausted battery. If you want to monitor the battery charge accurately, you need a digital panel voltmeter; you'll find them on eBay for less than £10. Wire it into the power socket on the switch panel. Charge the battery whenever you see less than 13.5 volts displayed. Many car's today with a three brush dynamo, have had all, or some of the bulbs changed to modern LED

replacements. LED's output more light and draw very little current compared to the incandescent bulb. Hardly any energy is wasted producing heat, which is why they give such superior performance.

## Charging faults

Low, or no output from the dynamo, is usually caused by poor commutator maintenance. Annually, you should remove the cover on the end of the dynamo and clean the commutator with a petrol soaked rag. Check that all three brushes are free in their holders and are not worn down to the extent that they are barely higher than the holders. If in doubt, remove the dynamo to the bench and remove the brushes. Clean the commutator with abrasive paper to a bright brass finish. Check with a fingernail, that the mica insulators between the brass strips, are not standing proud. If they are, they need to be cut back using a piece

*Digital voltmeter underneath the bulb horn. This has been permanently fixed to a 12v power plug, the socket of which is concealed under the dash. Consequently the voltmeter, can be unplugged if the car is 'on show'. Note the reading of 12.1v, indicating that the battery is very low and needs charging.*

of hacksaw blade. I would recommend also replacing the springs that bear down on the brushes. The third brush dynamo appears to need firm spring pressure to output correctly. A quick spring test if the problem is low or erratic charging, is to remove the commutator cover and run the engine, on charge, at a high idle. Then lightly press the brushes onto the spinning commutator. If a charge then registers, weak springs are your problem. A temporary cure, is to increase spring pressure by twisting in the correct direction, while holding the spring centre static, with a pair of thin nosed pliers. The correct dynamo brush spring pressure specification is 12 ounces +/- 2 ounces, for all three brushes.

Intermittent charging and an oscillating ammeter needle, is sometimes due to corroded contact points in the bulkhead mounted cut-out. Clean the points with fine grade abrasive paper and re-set them to 08 to 10 thou. Note, that to preserve these points, it is good practice not to leave the charge switch on, in stop start traffic. If you do, the points will be opening and closing a lot, together with associated arcing at the points.

If you strip the dynamo down, note that the nuts at each end of the armature are both right hand thread and are both 3/8" BSF. The two bearings in the dynamo that are lubricated via the oil nipples, are often over oiled. This oil subsequently getting onto the commutator and causing problems. Both bearings can be replaced with modern sealed-for-life types - they are part number 6302-2RSH SKF and available for less than £5 each from: www.bearing-king.co.uk

If your dynamo needs a complete re-wind and re-build, you are advised to contact Wood Auto Supplies in Huddersfield, who are specialists. The commutator is a hand built item and was made by RR themselves at Derby. When the old armature windings are removed, the commutator disintegrates in 90% of cases. Wood's rebuilds, include a hand made new commutator. You will receive what is in effect, a brand new 'old' dynamo. But it's not cheap at around £750.

Lastly, make very sure that the field fuse (No 2 in the fuse box) is a 3 amp type. The Handbook warns, *never to run the engine with the battery disconnected*. If you do forget, the 3 amp fuse should blow. A larger capacity fuse *will melt the field coil* and destroy the dynamo.

To give your dynamo an easy life and more importantly, to achieve an easy engine start, no matter how long the car has been standing, buy a Battery Optimiser. This is the new technology version of the old fashioned trickle charger. 'Accumate' is the leading brand. It will automatically maintain the battery in perfect condition and will also prolong its life. It works, by monitoring battery voltage. At a pre determined voltage drop, it gently discharges the battery further and then charges back to the optimum. The battery gets 'used' every day and stays 100% efficient.

In today's traffic conditions, modern flashing indicators on a vintage car are essential. Some owner's adapt existing lights to incorporate flashers, on the grounds that they do not want to change the appearance of the car. At the rear, this solution rarely works well, in terms of being seen. I believe it is safer to fit dedicated flashers at the rear. These do not have to be 'in your face'. SVC offer a really neat little pod flasher, finished in either chrome or gloss black. Nearly all cars have a rear bumper and all that is necessary is to fabricate a couple of sheet metal brackets, to fit them on. Cabling to the exposed rear indicators should either be in armour cabling, or to use original style metal conduit available from Fiennes.

Front indicators, are more difficult, as I would not want to spoil a wing by mounting an indicator. There are two

*Neat pod flashing indicator fitted to rear quarter bumper on GXO71.*

acceptable solutions. If your car has the older sidelights that look like mini headlamps, then it is relatively easy to replace the bulb holder with a twin filament variety, as used on stop & tail lights. The downside is that the indicator will flash white rather than amber. If your car has torpedo type side lights, then SVC offer a neat conversion. It replaces the single bulb holder with twin holders. One for a white sidelight bulb and the second using a unique high wattage amber bulb. These conversions work really well and show the correct flashing colour to the front.

Connecting to a 12v positive supply, will vary from car to car, as some cars will have been rewired and a spare fuse position will normally have been provided. GXO 71 had not been rewired and I decided the safest, most unobtrusive way, was to tap directly from the battery master switch. The cable connected to the switched side of the

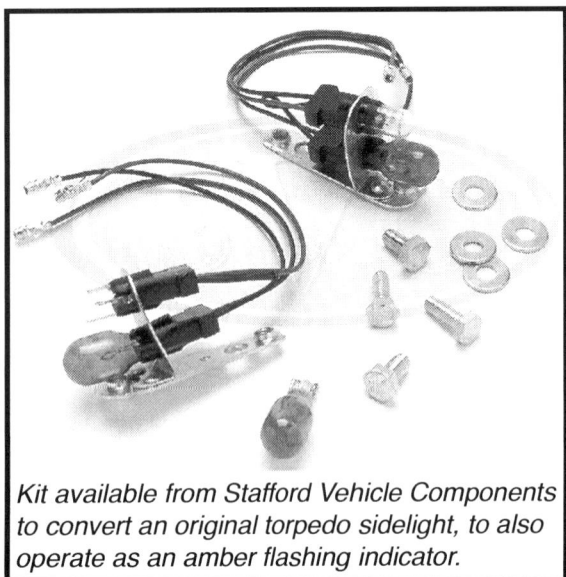

*Kit available from Stafford Vehicle Components to convert an original torpedo sidelight, to also operate as an amber flashing indicator.*

*Electrical additions on my 25/30. The panel is made from sheet metal painted gloss black and is screwed to the rear of the dashboard. From L to R: Power Socket for SatNav, Illuminated toggle switch for the Overdrive and an illuminated three way toggle switch for the flashing indicators.*

master switch, spurs to an in-line 10amp fuse, screwed to the side of the battery box. From the fuse, a cable feeds the overdrive unit and the flashing indicators. All the new cabling runs inside black plastic 'slit' conduit, fixed to the outside of the original conduit on the chassis.

See the back of this book for suppliers of a range of indicator switches and the flashing indicator unit itself. These come in all shapes and sizes. You need to choose one such as the Lucas SFB105 (42 watt) or the Lucas SFB 193 (32 watt). The wattage needs to match the wattage of the combined front and rear indicators. If there is a mismatch, then the flashing time period will be incorrect.

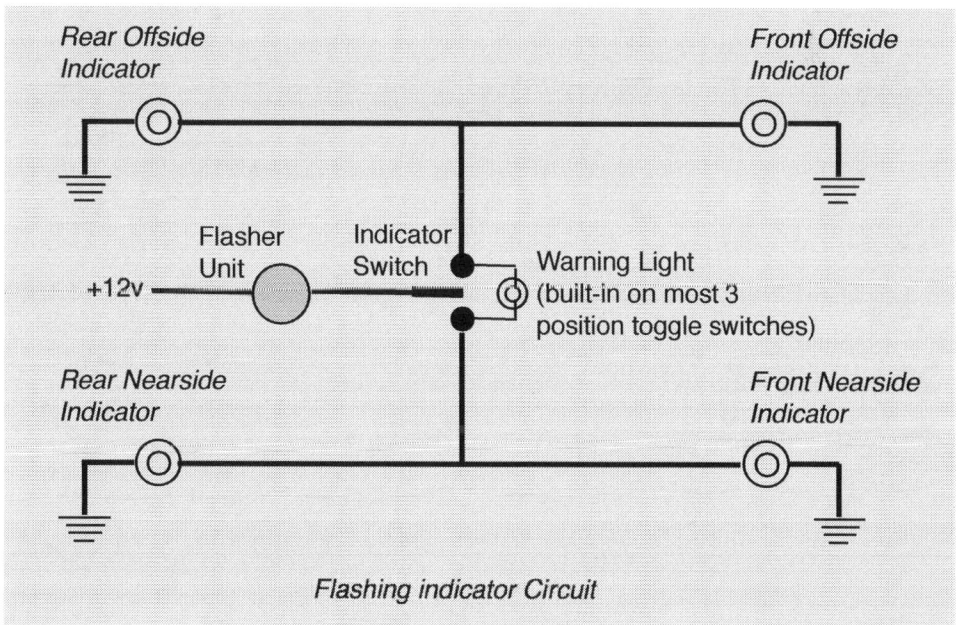

*Flashing indicator Circuit*

89

## GXO 71 Original wiring diagram

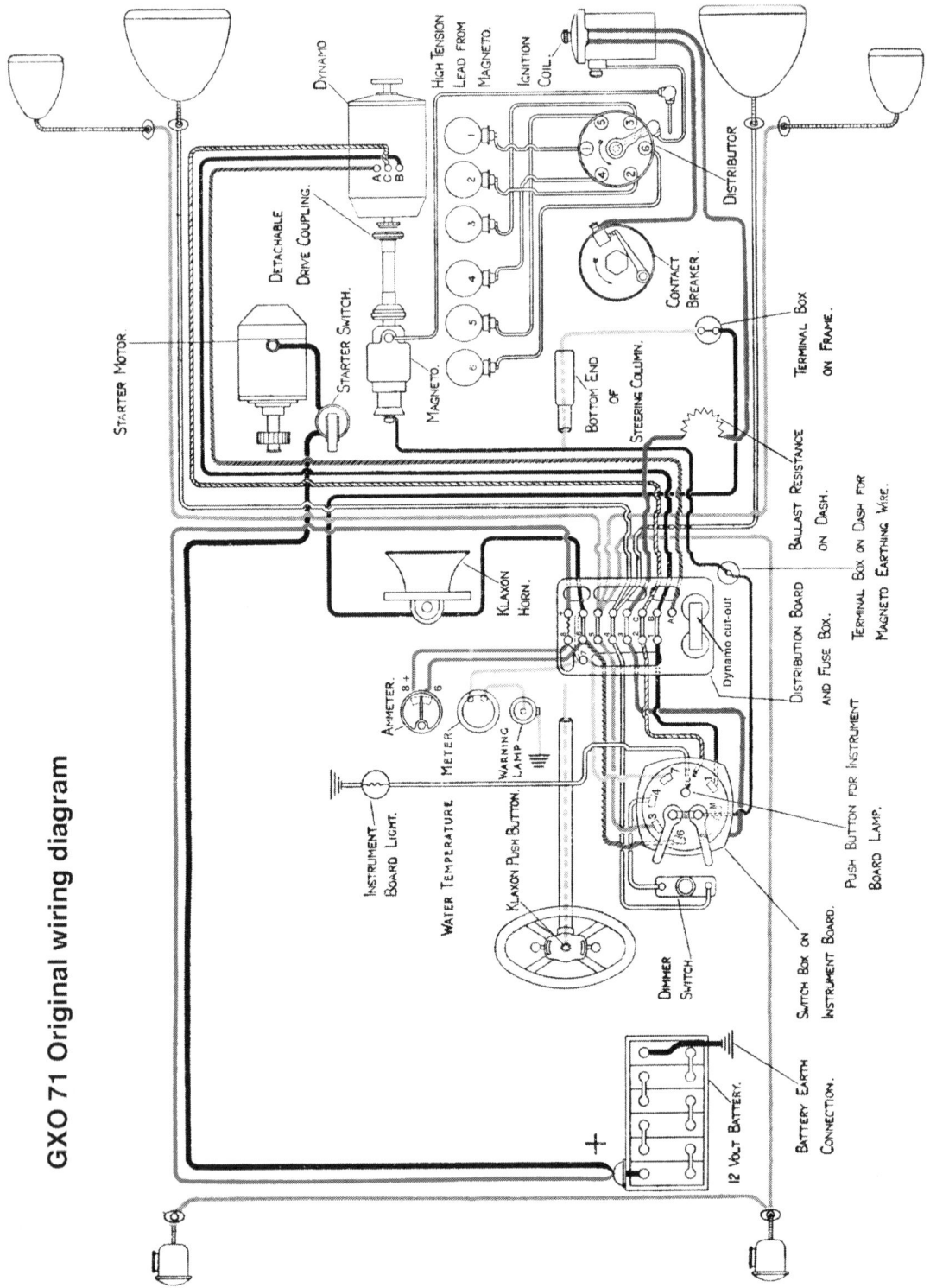

The braking system employed on the small chassis is quite ingenious. For years, Rolls-Royce cars did not have front wheel brakes. This was partly because Henry Royce thought powerful brakes on the rear wheels was sufficient, and also because in the days of poor tyre technology and unmade roads, a front wheel skid brought terror into the heart of the automobilist. However better roads and increased performance from the cars, dictated that all four wheels now needed to be braked. Henry's answer to this, was to adopt and then improve, a servo system first used on the Hispano Suiza chassis.

Imagine if you will, the action of a clutch in a manual car. As you increase the throttle and gently let the clutch out, the metal driven disk bolted to the crankshaft, is moved progressively into contact with a fibre disk connected to the road wheels.

*Rear brake adjustment is needed when the actuating cable slack exceeds 1.5" for the foot brake and 1.75" for the hand brake; minimum slack is .75" for both foot brake and hand brake.*

*To adjust: the rods nearest the road wheel actuate the foot brake. So remove pin **A** from the jaw **B**. The pin is held by a collar and split cotter. Slacken the nut **C** and screw the jaw further onto the rod as necessary. Adjust the other rear wheel by exactly the same amount.*

*Before replacing pin **A**, check hand brake adjustment, as it is not possible to remove jaw **D** when the foot brake jaw is in place. Adjust in exactly the same way, removing pin **D** from the inner rod. After adjustment tighten nuts **E** and **C** and fit new cotter pins to **A** and **D**. When no more adjustment is possible, the brake shoes need relining.*

Initially the two disks slip, the powered one gradually transferring its rotary motion to the road wheels. This transfer of power causes the stationery car, to start moving. Now think of a similar clutch arrangement, but this time feeding power not to the road wheels, but to pull on a rod, connected to the brakes. In a stroke, you have a powered braking system. This is essentially what the servo does. Except the power does not come directly from the engine, but from the gearbox. This design feature ensures that you have powered brakes whenever the car is moving - even if you are coasting down a hill with a dead engine, or the car is being towed. This design peculiarity needs to be borne in mind, if for instance you are parking on a hill. The car needs to be moving at +4mph to activate the servo; below this speed, braking efficiency is much reduced. It can catch you out, if your garage is at the end of a short downhill drive, where you might enter the garage and find you cannot stop before hitting the garage wall! Part of the reason for

*Front brake adjustment is indicated, when lever **F** can be moved by hand 7/8"
or more; ideally it should move by no more than half an inch.
To adjust: remove split cotter from nut **G** & remove. This will expose the
serrated adjustment; mark its position relative to **I** before removing. Unscrew **J**
enough to allow the serrated member **I** to be moved clear of serration's on
lever **F**. These 2 sets of teeth are marked with an arrow and the figures 0 to 5.
If the brakes have never been adjusted before, this figure will be 0. Tap lever **F**
away from the wheel and carrying with it serrated member **I**. Disengage the
serration's. Turn the cam operating shaft and with it **I** by using a spanner on **K**,
until the parts can be re-engaged to the next higher figure. Re-assemble in
reverse order.*

reduced braking below 4mph, is because the front brakes do not operate at all. The front brakes are not connected to the brake pedal, but to the servo. This is quite clever, because if the car develops a rear wheel skid, the rear wheels will lock up, which will cause the gearbox shaft to stop revolving, the servo therefore ceases to operate and the front brakes are released. Thereby ensuring the car continues to steer as the driver corrects the skid.

Another safety feature is the handbrake. Unlike other cars, the handbrake is *completely* independent. It operates on the rear wheels, where there are extra wide brake drums, each containing an entirely separate set of brake shoes, actuated by the handbrake. Hauling on the handbrake, therefore adds extra retardation. It also means that in the almost impossible event of the foot brake failing, the handbrake allows the car to be braked to a standstill, swiftly and safely.

Repairing faults in the braking system is not for the faint hearted and special tools are required to remove the hubs. There is an inexpensive manual available from the RREC "Rolls-Royce Small Horsepower Brake Systems" by R Haynes. It is essential reading if you plan serious remedial work on the braking system. On my car the brakes clearly needed adjustment and this is well within the scope of the average owner; see the captions under the pictures. However adjustment did not wholly cure the problem. Closer examination revealed that

the brake ropes that actuate the brake shoes, had stretched. The length of these ropes, measured from the centre of the eyes on the fork ends should be 61.9 inches (rears) and 44.75 inches (fronts). The front ropes on my car were the correct length. But the rears had stretched in service and now measured 62.5 inches. The (length) tolerance specified by Rolls-Royce on the brake ropes is just 50 thou - so Bluebell was long overdue in having them replaced. It's quite a difficult operation replacing the wire rope to this degree of precision. So I sent the existing ropes still complete with fork ends to Fiennes. The method of securing the wire rope in the fork end, is to bend the rope around a wedge that fits in the fork ends. The fork ends where the rope enters are then heated and filled with soft solder.

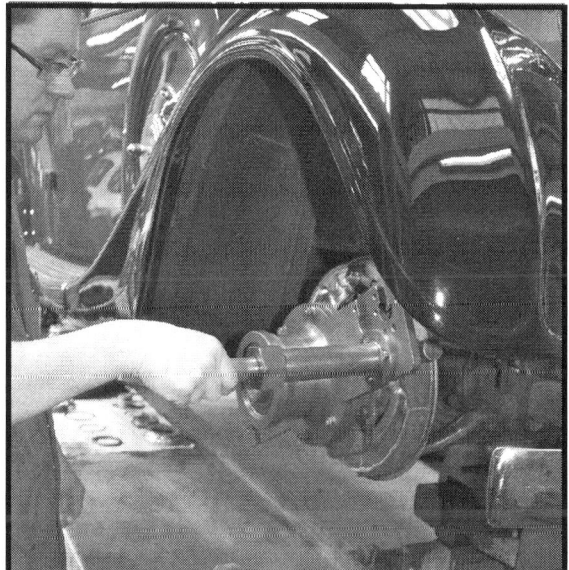

*One of many special tools needed to overhaul the brakes. This is a rotating cutter being used, to chamfer new brake shoes to precisely fit the brake drum. (25/30 at West Hoathly Garage)*

## Bumpers

Henry Royce did not approve of full sized bumpers, as he considered there was a possibility that they would twist the chassis in the event of a collision. However, many coach builder's fitted them. The front bumper on GXO71 had been removed and lost long ago and I decided not to fabricate a new one. The rear quarter bumpers had been incorrectly re-plated, using chrome instead of nickel. After stripping the chrome, one of the tubes was revealed to be badly corroded. So I had all four bumper tubular sections remade by Orchard Restorations in Sussex at a cost of £129.25. This work entailed machining eight new domed end caps in solid brass, as well as fabricating four new curved tubular sections in 3mm mild steel. The remade bumper sections were nickel plated and polished by Collonade at a cost of £199.75.

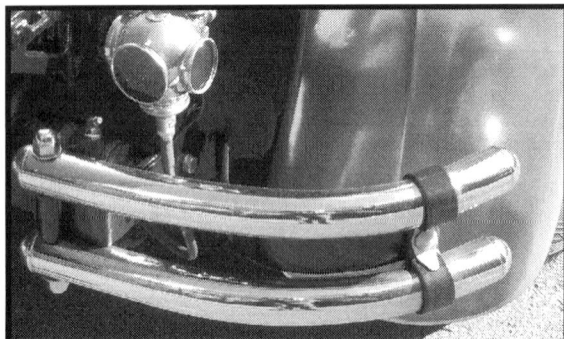

*The restored rear quarter bumpers.*

## Chassis lubrication system

The Bijur chassis lubrication system on the 20/25, consists of a pump and associated oil reservoir tank situated on the bulkhead, under the bonnet. The pump is operated by a foot pedal to the left of the steering column, in the front compartment.

In the handbook, there is an instruction to give one pump on the chassis lubrication pedal every 100 miles. You are advised to ignore this instruction, as the lubrication it provides is the minimum. When the car is first started, press the pedal fully down and drive off. As soon as the pedal returns to the top (1-2 minutes), depress again. Repeat this one more time, equating to three full pushes on the pedal. After that, depress the pedal once, every 50 miles. The above assumes the car is setting off on a longish run. The reason why the

Screw removed

*Front offside spring shackle with screw that seals the shackle oil-way drilling removed. Clean oil should ooze from here when the bulkhead foot-pump is operated.*

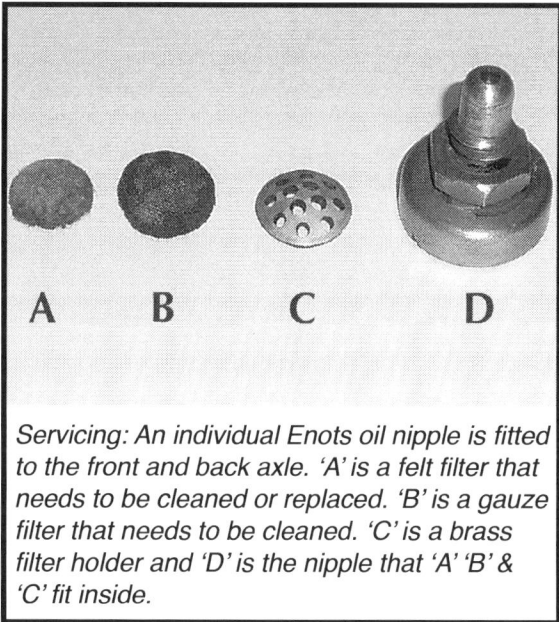

*Servicing: An individual Enots oil nipple is fitted to the front and back axle. 'A' is a felt filter that needs to be cleaned or replaced. 'B' is a gauze filter that needs to be cleaned. 'C' is a brass filter holder and 'D' is the nipple that 'A' 'B' & 'C' fit inside.*

care to be associated with cheaper cars of the period, that tended to drip oil as a matter of course. Today we are more concerned with the preservation and proper maintenance of our cars, than oil on the garage floor.

When the chassis pedal is depressed, pressurised oil is forced through a network of pipes, terminating in 'drip' valves, that feeds oil into various lubrication points around the chassis. Each drip valve, as the name suggests, 'drips' oil at a pre-determined rate. The oil flow is marked on one of the flats on each drip valve. 'S0, S1, S2' etc. The higher the number, the greater the flow of oil. This is a 'total loss' system,

original instruction is now considered to be not enough, is that when the cars were new, they were purchased by wealthy people. These people mostly lived in 'nice' houses. They did not expect the 'best car in the world' to leave oil deposits on their gravel drives. It is also the case, that Rolls-Royce did not

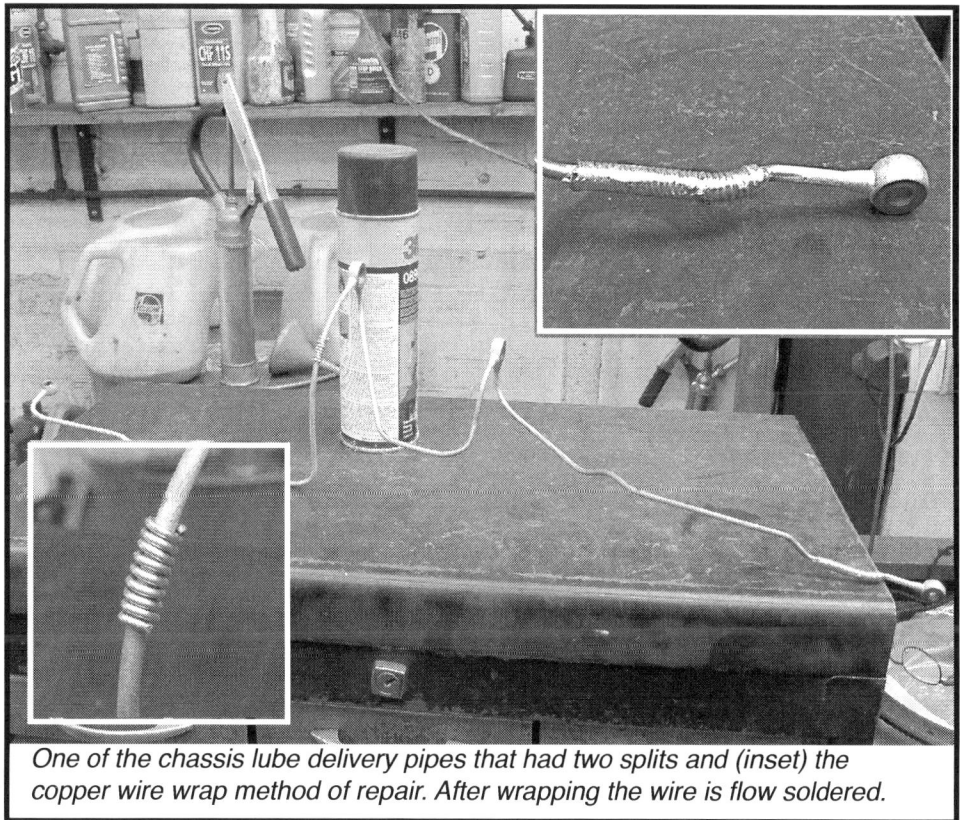

*One of the chassis lube delivery pipes that had two splits and (inset) the copper wire wrap method of repair. After wrapping the wire is flow soldered.*

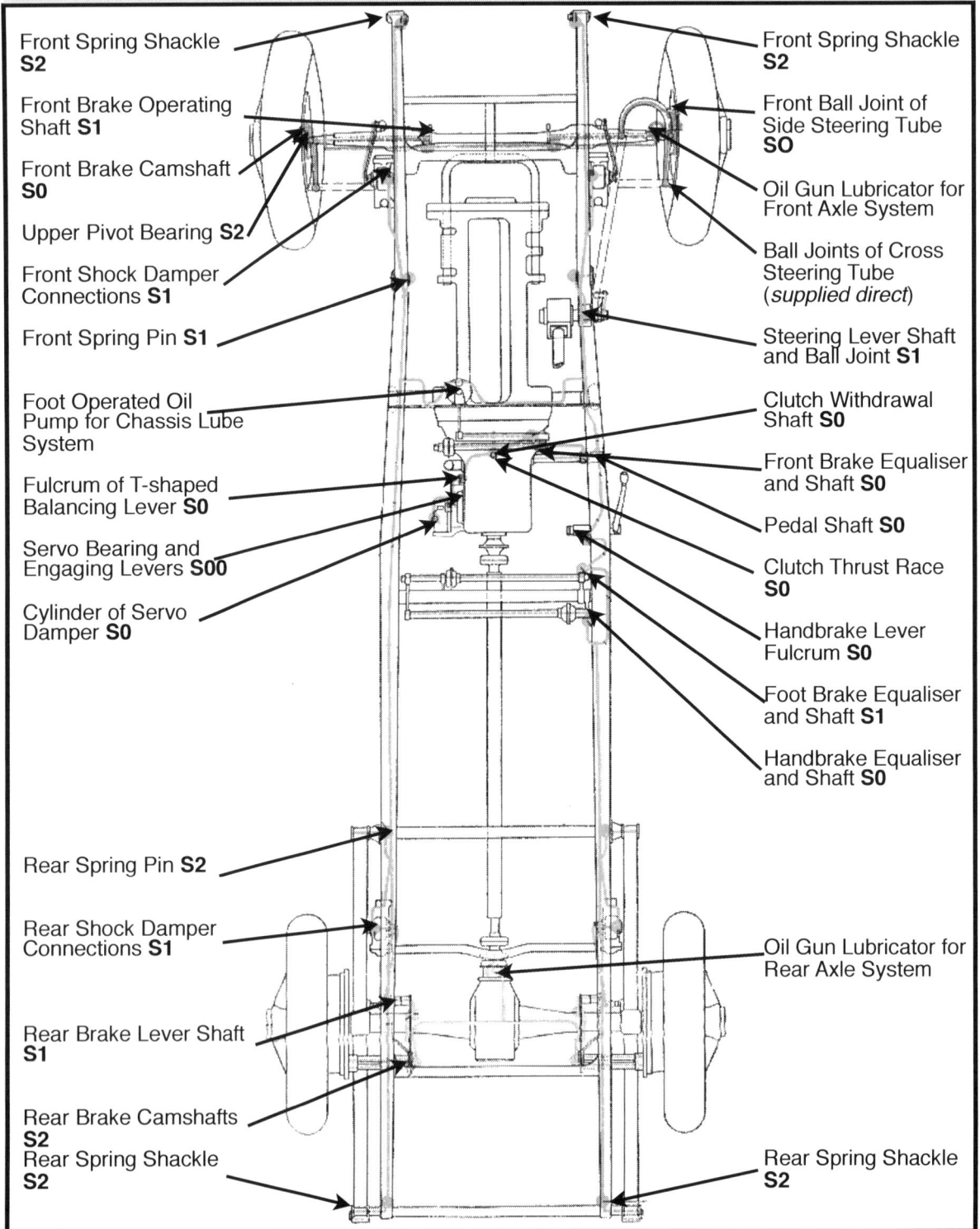

Front Spring Shackle **S2**

Front Brake Operating Shaft **S1**

Front Brake Camshaft **S0**

Upper Pivot Bearing **S2**

Front Shock Damper Connections **S1**

Front Spring Pin **S1**

Foot Operated Oil Pump for Chassis Lube System

Fulcrum of T-shaped Balancing Lever **S0**

Servo Bearing and Engaging Levers **S00**

Cylinder of Servo Damper **S0**

Front Spring Shackle **S2**

Front Ball Joint of Side Steering Tube **S0**

Oil Gun Lubricator for Front Axle System

Ball Joints of Cross Steering Tube (*supplied direct*)

Steering Lever Shaft and Ball Joint **S1**

Clutch Withdrawal Shaft **S0**

Front Brake Equaliser and Shaft **S0**

Pedal Shaft **S0**

Clutch Thrust Race **S0**

Handbrake Lever Fulcrum **S0**

Foot Brake Equaliser and Shaft **S1**

Handbrake Equaliser and Shaft **S0**

Rear Spring Pin **S2**

Rear Shock Damper Connections **S1**

Rear Brake Lever Shaft **S1**

Rear Brake Camshafts **S2**

Rear Spring Shackle **S2**

Oil Gun Lubricator for Rear Axle System

Rear Spring Shackle **S2**

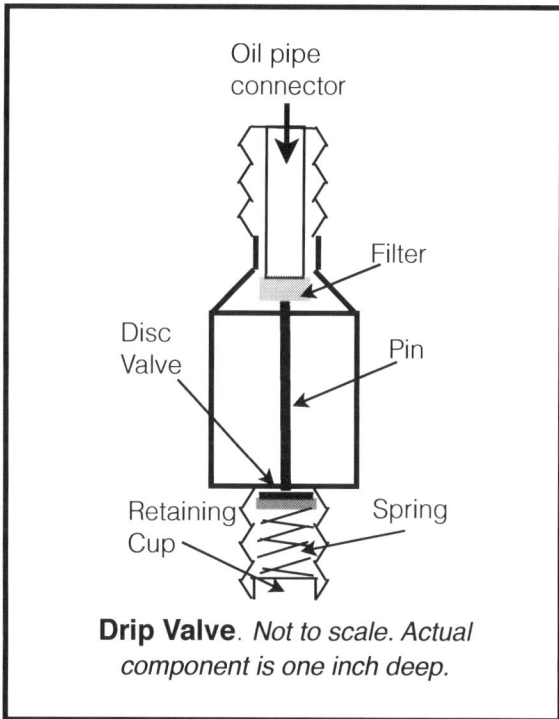

**Drip Valve**. *Not to scale. Actual component is one inch deep.*

The bulkhead pump is very robust and I've never heard of one failing. At the bottom of the pump is a nut. Every 20,000 miles undo this, to remove the cover. Collect and observe the position of the distance pieces and/or washers. Beneath these will be found a felt filter. This should be cleaned, or ideally replaced. At the same time, the filters under the Enots nipple on the front and rear axle should be cleaned, or replaced. See the picture in this section.

Failure of oil to exit from a particular drip valve, will probably be caused by a split, or blocked pipe. Or a blocked drip valve. For pipe repair, see the picture in this section. A section of pipe runs through the rear brake drums, so any sign of oil leakage here, indicates a broken pipe inside the brake drum. The factory advised against

which is why, if the system is functioning correctly, there will always be oil present on your garage floor. The system ensures that the various bearings it lubricates, last for a very long time indeed. However, known weak spots, includes the feed to the front spring shackles and where the oil goes on to lubricate the front springs. If the chassis lubrication system fails here, the bushes and pins in these shackles, which are exposed to road spray, will quickly wear and corrode. A quick check that all is well and which will also prove the integrity of the chassis oil pump, is to remove the 2BA screw from each of the shackles (see picture) and operate the one-shot foot pedal a couple of times. After a few minutes, clean oil should be seen to drip from the screw holes. Replace the 2BA screws and rejoice!

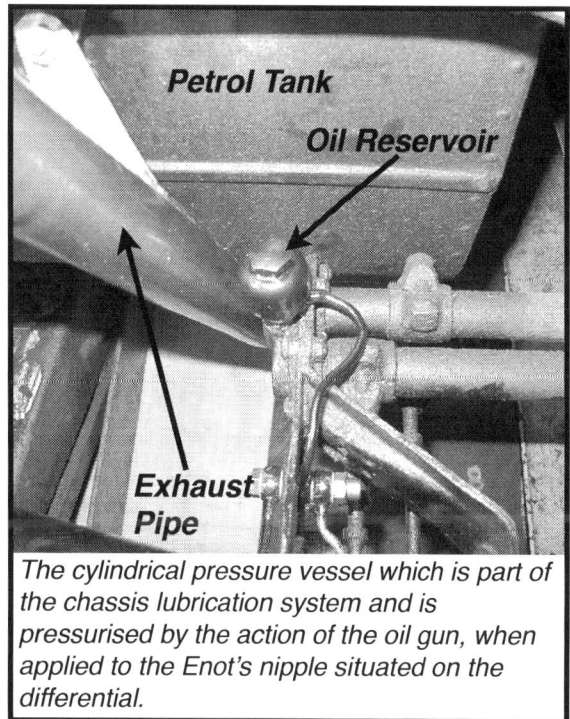

The cylindrical pressure vessel which is part of the chassis lubrication system and is pressurised by the action of the oil gun, when applied to the Enot's nipple situated on the differential.

trying to dismantle a drip valve. But it can be done, with patience. Use tweezers to remove the gauze filter from the inlet side. On the outlet side, use a scribe to prise the retaining cup loose. Be aware that the spring and gauze filter beneath it, will want to fly out. Perhaps dismantle the valve with it and your hands inside a polythene bag. With these parts removed, the pin valve can then be gently tapped out. Clean everything and re-assemble. You will probably have to peen the end of the plug, to hold the retaining cap in place. Drip valve spares are available from the original manufacturer, alternatively, buy a complete new drip valve from Fiennes. They're not expensive.

The chassis lubrication system does not extend to the front and rear axles. On the offside front of the front axle, you will find an Enots oil nipple. There is a similar nipple on top of the rear differential. Every 500 miles oil should be screwed in with the Enots gun until the handle goes tight. Wait for a minute and then repeat. Do this four times. The oil you have injected will fill and pressurise a container. Which in turn feeds oil to bearings within the axle systems. Do not neglect to do this regularly, as the front system feeds the oil hungry king pins. Over oiling is better than no oil! There is a known weakness in this area; the oil reaching the rear spring shackles is marginal at best. The work-around solution, is to fit oil nipples to the rear spring shackles and to inject oil with a gun, every 1,000 miles.

Finally and again contrary to the Handbook, do not use engine oil in the chassis lubrication tank. A modern EP80 oil should be used. These extreme pressure oils do a better job than engine oil, especially in areas such as the king pins.

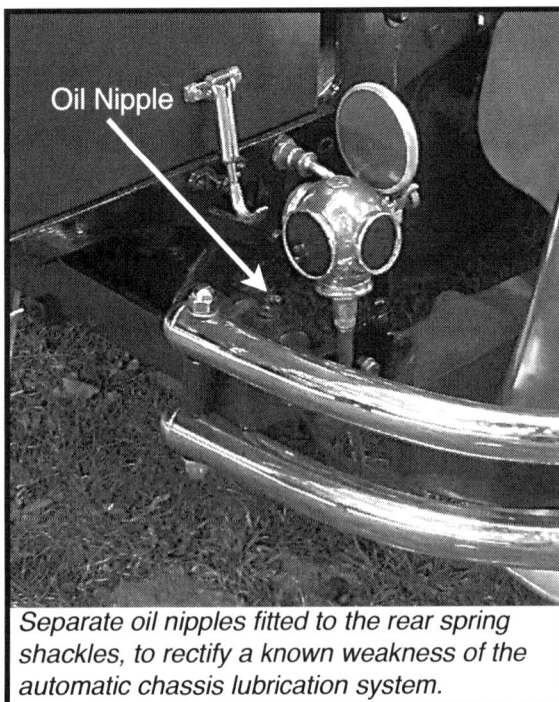

Oil Nipple

*Separate oil nipples fitted to the rear spring shackles, to rectify a known weakness of the automatic chassis lubrication system.*

Fitting an overdrive transforms these cars, as they were all to a greater or lesser degree, under geared. They had to be, because the factory had no real control over the weight of the body that would be fitted. Consequently the chassis was geared for the heaviest body envisaged, making the lighter bodied cars, especially the tourers, very under geared. With an overdrive, top speed is improved, but that is not the main benefit. That comes from a quieter, less stressed lower revving engine, with better fuel consumption. Engaging the overdrive, drops the revs by 22%, while at the same time, increasing the speed of the car by 28%. A win-win! The prospect of driving on a motorway is no longer to be feared, where with overdrive engaged, the car soon gets into a long legged stride. The engine of the four speed version of the 20HP and the 20/25 at 50mph, is turning over at circa

2,500rpm. Engaging overdrive at exactly the same throttle opening, will boost the speed to 64mph. The beauty of the overdrive, as opposed to fitting a higher ratio differential, is that when disengaged, the low speed 'chug-ability' of the car is unchanged.

The installation consists of a 'J Type' Laycock separate unit overdrive, as used to be fitted to Ford Transit vans etc. So it is rugged enough to cope with the torque of the Rolls-Royce engine. There are two ratios available, a 28% and a 25%. 28% is recommended and the two versions can be identified by a /8 (or/5) following the serial

*Typical overdrive installation on GXO71. The wiring is temporary at this stage.*

*The power supply to the Overdrive. The arrow marks the Fuse/holder, mounted on the side of the battery box.*

number on the casing. The unit is connected to the gearbox by a short prop shaft; which means that the existing prop shaft has either to be replaced, or cut down & re-balanced. At the bottom of the gear stick is fitted a limit switch. This ensures that the overdrive can only be engaged, when in top gear. This is because the hydraulic pressure required by the overdrive, is delivered by an internal pump that is driven from the overdrive output shaft. The lower speed when in a lower gear, would not produce enough pressure for the overdrive to work properly.

On the face of it, fitting the overdrive is a simple job. But it is extremely important to get the angle of the prop shafts correct and for everything to align correctly. Get it wrong and you will possibly have a mysterious vibration - even with the overdrive disengaged. You are advised to use a specialist kit, such as Nostalgia Cars

offer. The complete kit is currently £2,500 plus vat. This firm will also fit the unit if required, for an extra £600. All the vendors that advertise overdrive installations, obtain the main unit from Sheffield Overdrives. This firm can also repair or re-build the core unit.

### Servicing & common faults

Do not use anything other than red ATM fluid in the overdrive. Drain and re-fill annually with 1.25 pints of fluid.

If the (engaged) overdrive jumps in and out of operation as you drive, it is almost certainly due to the limit switch at the base of the gear lever coming loose.

If the overdrive will not engage, first check that power is getting to it and the solenoid is working. If you get under the car while an assistant switches the overdrive on (engine off and top gear selected) you will be able to hear the click of the solenoid as it actuates. Connect a multimeter in series with the positive cable going to the overdrive and measure the current when it is switched on. It should be 2 amps; it is normal for the solenoid to get hot in use. If the power is OK, then the cause of not engaging is probably muck underneath the relief ball valve. Drain the fluid and remove the sump & flat filter underneath. There are 3 plugs underneath the filter. Remove the middle one with a drain pan underneath to catch the ball, ball seat and spring. Clean all and re-assemble. You will need a pin wrench to remove the plug. Tighten to 16 lb/in when re-placing.

There are four shock absorbers bolted to the chassis, operated by a damper arm that is fixed by a link, to the road spring.

The shock absorber has two important features. It will not move in either direction, until the load on the damper arm is sufficient to lift one or other of the valves. In this respect, its behavior is similar to the earlier friction damper, which will not move until a certain poundage is applied to the arm. However, due to the larger size of the low pressure valve, it lifts against its spring loading easily. The smaller high pressure valve, demands a higher pressure to be applied, before it will lift. The shock absorber therefore, delivers a bigger resistance to rebound deflections than to bump deflections. Once the valves have lifted, oil flows past them by running along grooves cut in the valve stem. This

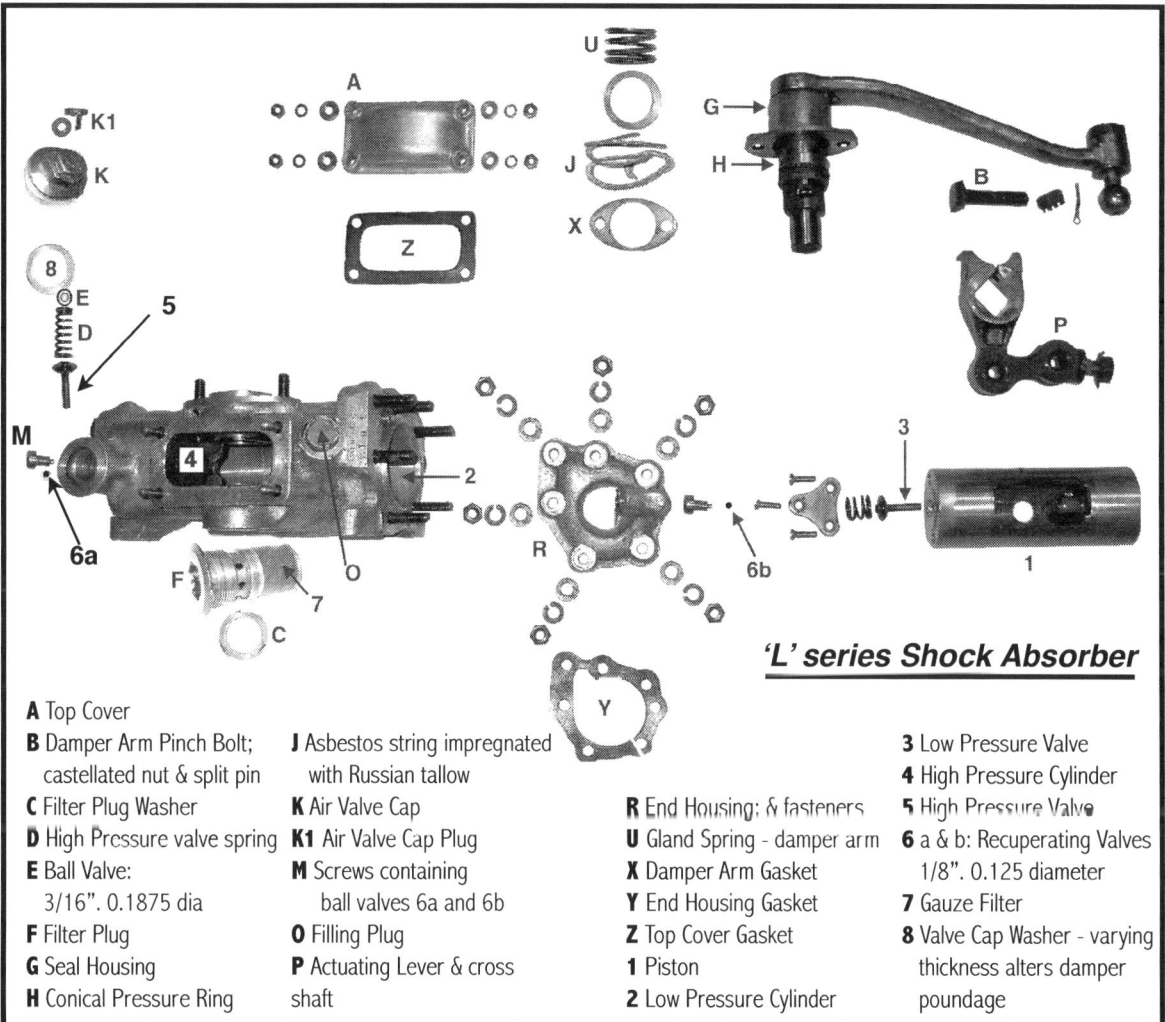

**'L' series Shock Absorber**

**A** Top Cover
**B** Damper Arm Pinch Bolt; castellated nut & split pin
**C** Filter Plug Washer
**D** High Pressure valve spring
**E** Ball Valve: 3/16". 0.1875 dia
**F** Filter Plug
**G** Seal Housing
**H** Conical Pressure Ring

**J** Asbestos string impregnated with Russian tallow
**K** Air Valve Cap
**K1** Air Valve Cap Plug
**M** Screws containing ball valves 6a and 6b
**O** Filling Plug
**P** Actuating Lever & cross shaft

**R** End Housing; & fasteners
**U** Gland Spring - damper arm
**X** Damper Arm Gasket
**Y** End Housing Gasket
**Z** Top Cover Gasket
**1** Piston
**2** Low Pressure Cylinder

**3** Low Pressure Valve
**4** High Pressure Cylinder
**5** High Pressure Valve
**6** a & b: Recuperating Valves 1/8". 0.125 diameter
**7** Gauze Filter
**8** Valve Cap Washer - varying thickness alters damper poundage

resistance, provides additional damping. The low pressure and high pressure valve is not adjustable on early cars. Later cars feature an adjustable high pressure valve. This is done by changing the thickness of the adjusting washer 8 under the valve cap. A thicker washer will soften the damping action and vice versa.

### Maintenance

This consists of scrupulously cleaning the filler plug and surround, on top of the shock absorber. Top-up, using either Mineral SAE20 oil (as sold for Motorcycle telescopic forks) or SAE30 Mineral, which will give a slightly stiffer action. If the existing oil looks discoloured, drain it off, by removing the filter plug 7 at the bottom. And, of course, wash the filter using neat petrol.

If one or other shock absorbers is suspect, then undo the damper arm and move it up and down by hand. You should be able to feel the resistance. Also check to see if there is any play in the arm itself. Problems inside the shock absorber are usually due to dirt and/or degradation of the valves and balls. The valves should be lapped onto their seats using metal polish and the balls replaced. Be careful when stripping a shock absorber, to ensure that the valves and springs go back in the same place. Be aware that the factory introduced a modification on the front shock

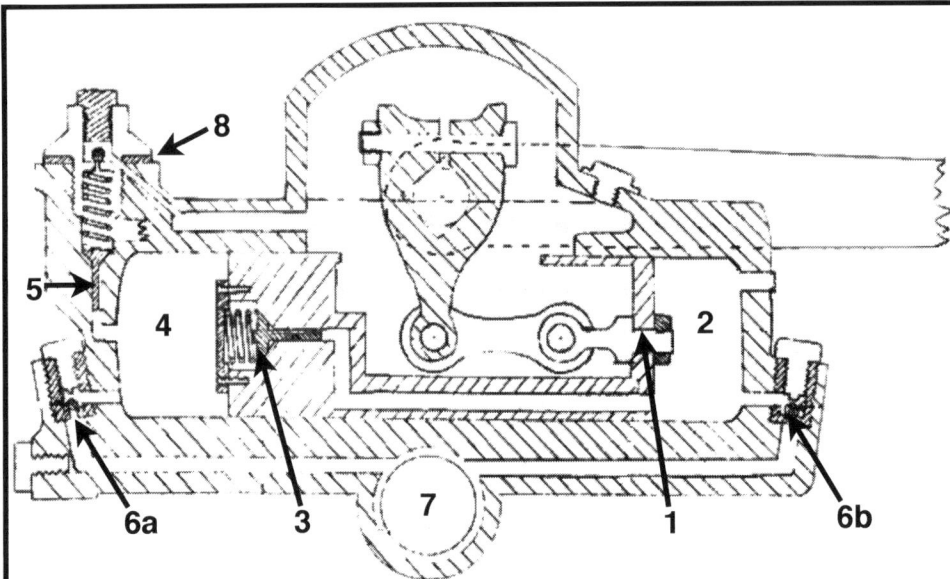

*When the wheel hits a bump in the road, the damper arm moves upwards, moving the piston 1 to the right. Oil in the low pressure cavity 2 is then forced by the piston through the low pressure valve 3 and into chamber 4. As the road wheel passes the bump, the damper arm moves down. This action forces the oil in high pressure cavity 4, past the high pressure valve 5. The oil is then returned to cavity 2. Oil that leaks past the piston into the central oil reservoir, is replenished, by oil entering the operating cylinders, through the recuperating valves 6a and 6b. These valves are fed with oil, after it has been filtered through the gauze filter 7. This filter connects to the main oil reservoir, remove it, to drain the shock absorber. A ball-controlled air bleed valve is incorporated in the valve cap at 8, to make the damper self-bleeding.*

absorbers, to address the problem of 'knock' on some cars. This was to drill a $1/32$" (.8mm) hole in the head of the high pressure valve - No 5 in the picture on the preceding page. You will see that there is a screw head in the top of this valve, this is to facilitate lapping the valve to its seat, when re-conditioning.

*Shock Absorber link. To adjust; remove leather gaiter (not shown). Slacken lock-nut A, tighten adjuster B until it nips up. Check that shaft can still be rotated a few degrees either way - the idea is to slightly pre-load the ball pins within. Tighten the lock nut, smear the whole shaft with copper grease and re-fit the leather gaiter.*

## Spring gaiters

Leather gaiters that lace around the four road springs, were fitted to all chassis. Each gaiter is fitted with a number of Enots oil nipples, so that the road springs operate inside a lubricated, virtually oil-proof casing. The springs were also cadmium plated with a dull polished finish. This combination of a fine finish and oil lubrication, gave superior and quiet suspension, compared with the average 'cart sprung' car of the period. It also meant that the shock absorbers had to be of a high standard, as there was no longer any significant friction within the springs, to damp out road shocks. Naturally, Henry designed and manufactured his own hydraulic shock absorbers.

On many cars, the leather gaiters have either perished, or have been removed altogether. They do need to be present, to maintain the high ride quality of these cars. The gaiters are still handmade to day by 'Wefco', the same firm that originally manufactured them. The Enots oil nipples, complete with mounting base for these gaiters, are available from Fiennes. They come complete with copper rivets, for holding the mounting plate to the leather. There are a total of 12 nipples on the gaiters - 2 at each side on the fronts and 4 at each side on the rears.

If you are re-furbishing existing gaiters, remove the waxed cord lacing. It is unlikely this will be able to be re-used. An acceptable substitute, is to use leather boot laces, or hand made leather laces from an equestrian supplier. Jack up the car and place blocks under the chassis rails, so that the wheels and axle hang free. The weight of the wheels and axle will pull the road springs down and open up the leaves. Then scrub the leaves with a stiff brush, using paraffin or white spirit. When they're nice and clean, paint all the leaves with EP 90 mineral oil. Use a brush to get the oil right down between the leaves. Using standard medical bandage available from any chemist, bandage the complete road spring. Use a stiff brush to stipple more oil onto the bandage. Then lace the gaiter in place and finish by injecting more oil via the oil nipples.

Annual maintenance consists of injecting a full Enots oil gun of EP90 into each nipple, with paper spread under the car. Leave overnight to let the oil drip; you will find pools of dirty oil under the springs in the morning. This job can usefully be combined with cleaning and greasing the wheel hubs, as with the wheels removed, it is easier to get at all the nipples. If your car has lost all its gaiters and you're looking for a cheaper solution than new leather gaiters, then use 'Denso' tape. This is a waxed tape available from Plumber's merchants and used for protecting pipes that are to be buried. Clean and oil the springs and then wrap them in the tape. Over a short period of time, the tape will meld itself into one. Nothing like as efficient, or as elegant as Wefco gaiters, but a whole lot better than leaving the springs naked.

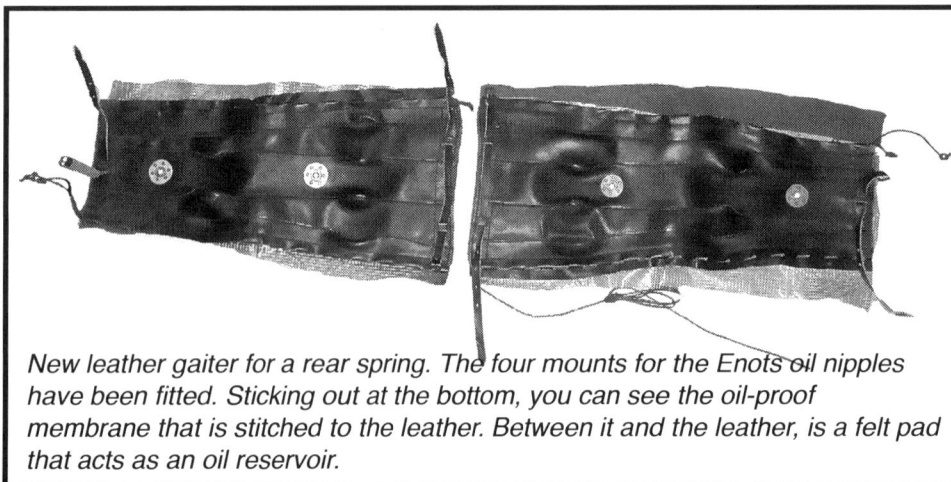

*New leather gaiter for a rear spring. The four mounts for the Enots oil nipples have been fitted. Sticking out at the bottom, you can see the oil-proof membrane that is stitched to the leather. Between it and the leather, is a felt pad that acts as an oil reservoir.*

Some cars suffer from quite severe steering wheel shake, known colloquially as the 'jaggers'. This is nearly always due to incorrect tracking; although it can also be due to out of balance wheels. On GXO71 I experienced rapid tyre wear at the front, but apart from occasional mild steering wheel shake, there was little to indicate a problem. When I replaced the tyres, I got the tyre shop to check the front wheel toe-in using their accurate laser equipment. This showed it to be 20mm. The toe-in stated by Rolls-Royce should be in the range zero to five sixteenths of an inch. No wonder the tyres wore out! Adjusting the tracking is a simple procedure and is outlined in the picture caption. Note that on earlier cars, the method of adjusting toe-in, is to bend the cross steering lever. This has

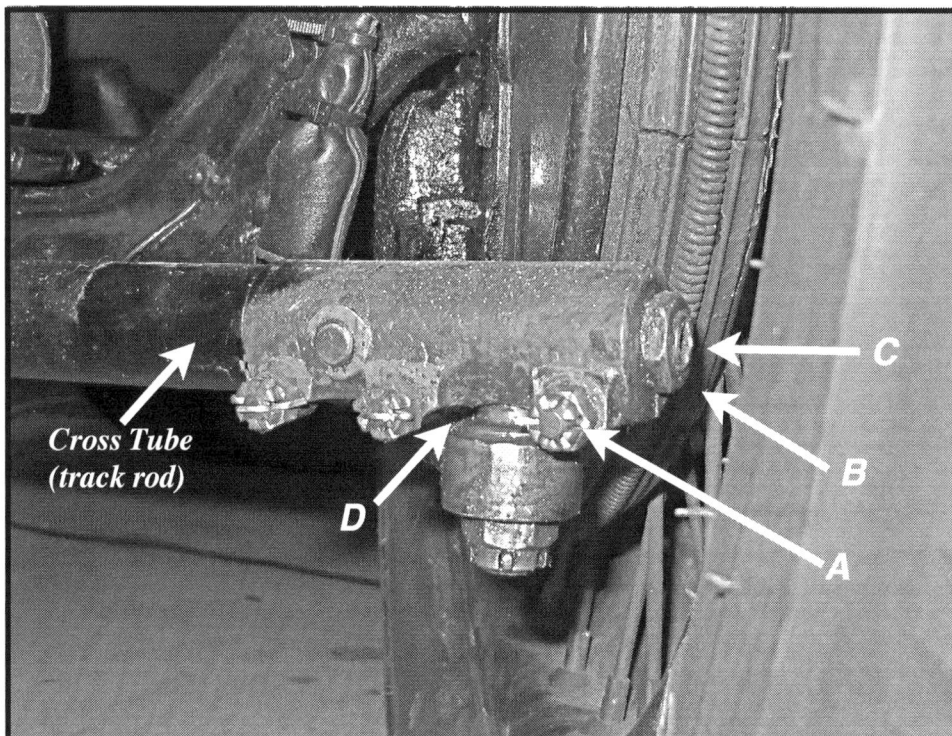

*The offside cross steering tube ball joint with the leather gaiter removed. To adjust: Remove split pin and slacken 'A'. Slacken lock-nut 'B'. Using a 6mm square key, slacken 'C' by at least one full turn. This will now release the spring pressure on the ball pin and the cross tube should be easy to rotate within the arc allowed by the ball. Move to the nearside ball joint and repeat the above. This time tighten 'C' after inserting a .0005 - .015 feeler gauge into the gap you will see if you look directly up underneath the ball joint at 'D'. This is the joint between the steering tube and the brass ball-pin carrier and it is important that a small gap is left here. When finished it should be possible to just rotate the cross steering tube using one hand. Go back to the offside and repeat this adjustment. When finished it should be possible to just rotate the steering tube, using both hands. The adjusted ball pins as outlined above, introduce enough friction into the steering, to avoid steering wheel shake when driving on a rough road. The adjustment also physically moves the ball pins to achieve the original factory set front wheel toe-in, of zero to a maximum of 9mm.*

the effect of moving the ball on the ball joint in or out very slightly. It sounds brutal, but that is what the factory did. Another way of achieving the same end, is to get a machine shop to make up shims to go on the brass pieces which contacts the ball. If you do need to go down this route with a machine shop, you are advised to spend £10 and get the relevant factory drawing from the RREC.

## Wheels

The most important annual maintenance regarding the road wheels, is to remove the disks (if fitted) and to check the spokes for tightness. Also check that there is no evidence of rust forming around the spoke heads, where they enter the rim. If there is, or there are broken spokes, you should consider a wheel re-build by a specialist, such as MWS. It is also essential to grease the splines that the wheel fits on,

at least annually. This is mainly to stop the wheel fretting and wearing the splines.

When putting a wheel back onto the car, fit the wheel spanner to the central nut, making sure that the two 'lip's actuated by levers on the spanner, are behind the rim of the nut. Moderately tighten. Lower the car off the jack. Tighten the central tommy bar on the wheel spanner so that the sprung red portion in the middle of the wheel nut is fully depressed. Then fully tighten the wheel nut; the last half turn or so by hitting the spanner with the mallet provided. Release the central fastening on the spanner. If the red inner part does not spring out flush with the wheel nut, then without

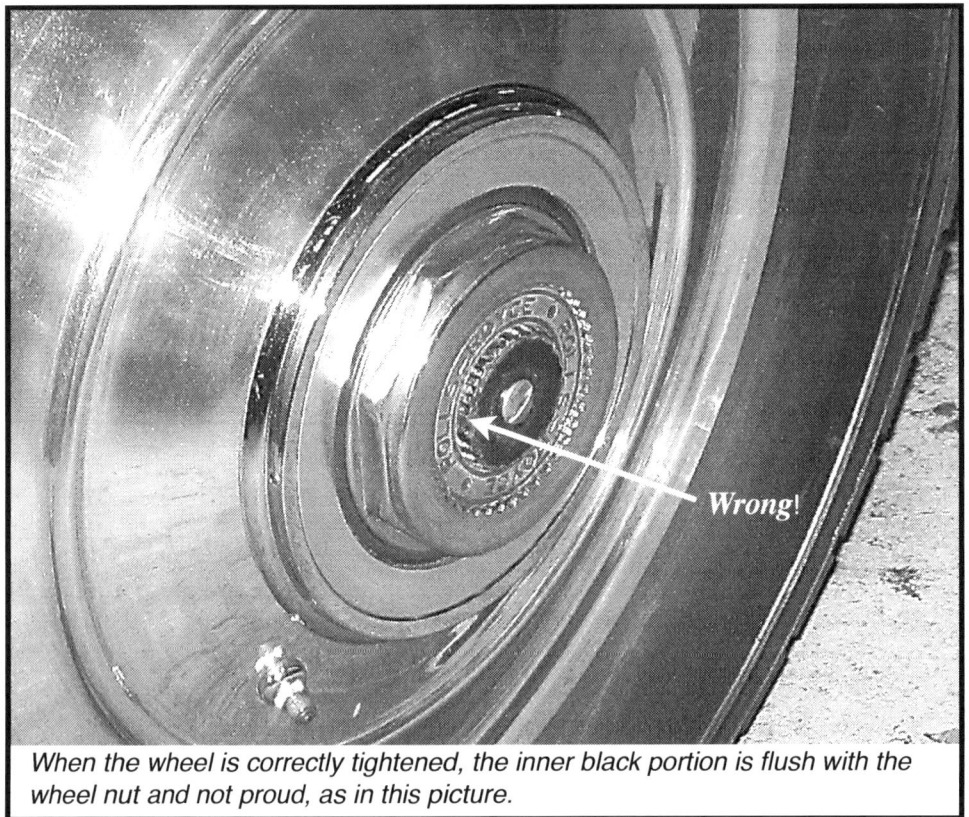

*Wrong!*

When the wheel is correctly tightened, the inner black portion is flush with the wheel nut and not proud, as in this picture.

tightening the central fastening again, tap the spanner to align the teeth on the cogs. This will release the torque and allow the central red part to move fully out. You may think I've laboured this operation, but it is surprising the number of cars I've seen with the wheels incorrectly tightened.

On later cars, the large nut that holds the wheel will come off, when the wheel is removed. On earlier cars the nut will stay on the wheel, where it is held by a circlip. If you need to blast and paint the wheel, or wish to re-nickel the nut, then the nut needs to be removed. This is really fiddly. Firstly, the wheel hub has to be spotlessly clean internally, to stand a chance of seeing the ends of the large circlip that holds the nut in place on the wheel. The wheel (with tyre still fitted) needs to rest face down on a spare tyre, to create a space underneath for the wheel nut to fall into. Then locate the circlip (looking through the wheel hub) and using circlip pliers, open the ring, so that you can get a long pick between the circlip & hub and then work the pick around. Easier to say than to do! Re-fitting is easy, as the circlip just snaps back into the groove.

### Painting a wheel

The original finish on the wheels was stove enamel. It is my opinion, that a correctly applied baked two pack finish is superior. But you do need to use a specialist wheel shop, as some owners have had very disappointing results, using two pack applied by a general paint shop. What tends to happen, is that the paint cracks around the spoke heads allowing moisture in and rust to form. The wheels on my cars were all taken back to clean metal and refinished by 'Lepsons'. The typical cost is, £55 +vat per wheel.

### Balancing the wheels

You will note the three stacks of weights, evenly spaced around the inside rim. On later cars, the weights are concealed

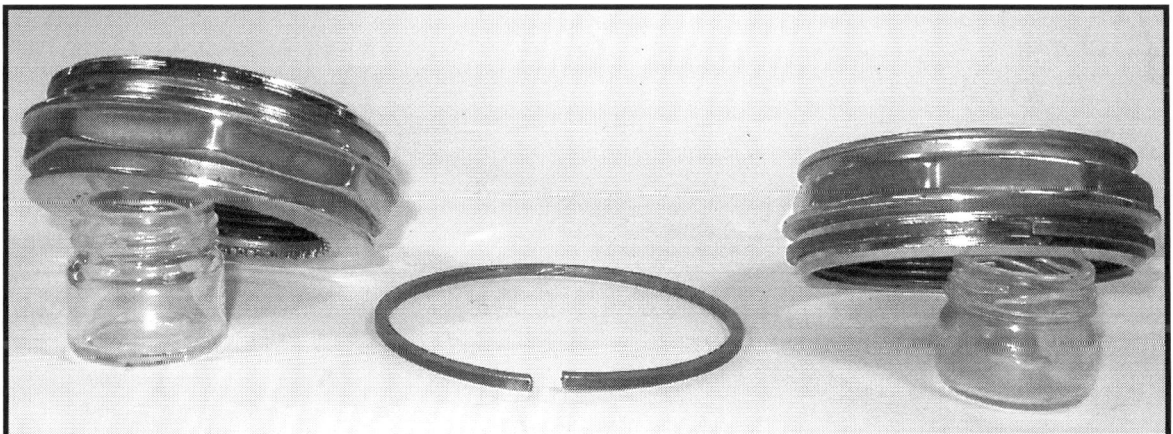

*Wheel nuts. The circlip from the nut on the left has been removed. The nut on the right, has the circlip in position, as it would be when the nut is on the hub. Note this nut is pictured upside down. These wheel nuts are easy to fit to the wheel - but very fiddly to remove.*

inside covers. Early cars did not have covers and the lead balance weights are usually painted the same colour as the wheel. The wheel & tyre is balanced by means of these circular lead weights. The weights are easy to make, using builder's lead flashing 1.2mm thick. Each finished lead disk will weigh 15 grammes. If your car has the 'cotton reel' covers, then use one to score around onto the lead. The lead can then be cut using stout scissors, just inside the scored line, to achieve a snug fit within the cover. If it is the case, that few weights are needed on any given stud, then the factory fitted lightweight bakelite disks, to bring the total lead/bakelite disks up to the height required, to secure with the top nut. Make a new cork gasket to go between the wheel rim and the first weight. Mount the wheel (with tyre fitted but minus wheel disk) onto a front hub. that has been checked that the brake is not binding. Taking the valve as a datum point, position the valve at 12

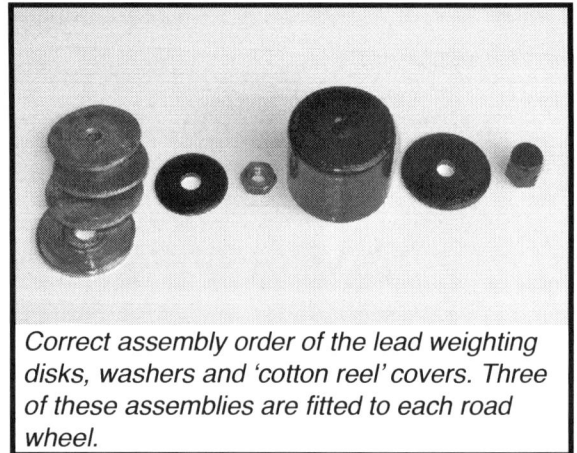

*Correct assembly order of the lead weighting disks, washers and 'cotton reel' covers. Three of these assemblies are fitted to each road wheel.*

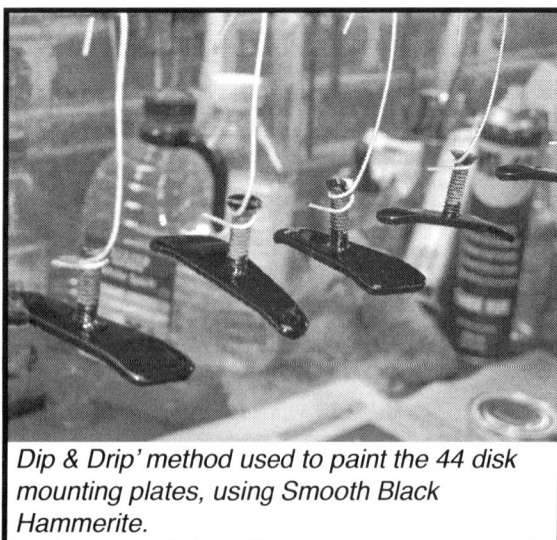

*Dip & Drip' method used to paint the 44 disk mounting plates, using Smooth Black Hammerite.*

o'clock, hold the wheel still and steady, then gently release. If the wheel rotates, add a couple of lead disks to the 'light' side and run a quarter inch BSF nut down finger tight. Add a nut to the two other naked studs, to cancel the weight of the nuts out of the equation. Repeat the test at 12 o'clock and add more lead disks as required. Continue by positioning the wheel at 3 o'clock and adding lead weights as necessary. Then the same for 6 o'clock and 9 o'clock. The aim is to finally be able to position the wheel at any point on the clock face, hold it motionless , let go and observe zero wheel rotation.

If a wheel is so 'bad' that you cannot balance it using the original lead weight method, then you might like to try centripetal balancing. This is the accepted method used by trucking and other commercial firms. You need to source 'zircon' ceramic beads small enough to pass through the valve and into the inner tube. You will need 142 grammes of beads for each wheel. Pour the beads in and that's it.

*Injecting 5 ounces of ceramic zircon beads, to achieve 100% perfect balance.*

## Wheel Disks

New aluminium wheel disks to fit a standard wheel are available from Fiennes Engineering. A wider range is available from LaMarr in the USA. Be aware that some 20/25 chassis, including GXO71, were fitted with heavy duty 'Colonial' wheels. These wheels have two sets of spokes that radiate from the outer and centre of the wheel rim and a third set of strengthening spokes radiating from the inner wheel rim; standard 20/25HP wheels have just two sets of spokes. This extra set of spokes, sets the wheel hub out at a wider angle to the standard wheel. Therefore standard wheel disks will not fit. I ordered a set of six (two spare wheels are fitted on GXO71) custom made disks from Fiennes. These were not cheap! Together with all the necessary fittings, they cost £1731. Note that Thrupp & Maberly fitted a single wheel disk to the outer of the wheel. Other coach builder's, notably Barker, also fitted a disk

The beads work in the same way that washing does, when first put into a spin dryer. The spin dryer drum wobbles around out of balance, but as the speed rises, the clothes distribute themselves within the drum, under the influence of centripetal force and the spin smoothes out. The beads work in exactly the same manner. The superiority of this method, is that the beads re balance the wheel every time the vehicle exceeds 20mph or so.

*The various components that make up one wheel disk.*

to the inside of the wheel as well. Henry Royce did not care for disks. He considered they added unnecessary unsprung weight and also increased the noise level inside the car. Owner's, tired of the work involved in cleaning a spoked wheel, liked them very much indeed!

## Clock & Speedometer

The Jaeger chronometric (clockwork) speedometer was incomplete when I acquired the car, in as much as the trip recorder knob and drive was missing. The matching Jaeger clock was also faulty and stopped and started at will. Both instruments were removed and sent to Vintage Restorations in Tunbridge Wells.

Apart from the missing trip knob, the speedo was found to be in good order. A replacement trip drive and knob was fitted and the instrument lubricated. The clock was in a much sorrier state and a new internal pivot and some bearings had to be replaced. Total cost £119.85 Both instruments now work correctly.

*Jaeger clock & chronometric speedometer (dashboard shown before restoration).*

The commonest door fit problem, is that of a door 'dropping', usually evidenced by scuff marks on the sill or body underneath the door. By rocking an open door up and down and observing the hinges, it will be apparent whether the vertical play in the door is caused by wear in the hinges, or whether the hinges themselves, are moving in the door or in the body. If the problem is the hinges, then they need to come off and either be replaced or repaired. If the movement is in the door or body, then this is either frame damage, or more usually rot within the wood surrounding the screws.

If it is frame damage, then self evidently, the car will need the attention of a bodyshop. Rot within the screw holes in the wood, can often be sorted, by applying a propriety fungicidal treatment. Then hammer into the screw holes a plastic wall plug, enlarging the holes slightly if needs be.

If the door is warped, then a 'straightener' is called for. Remove the door trim and window winder mechanism, if fitted. The straightener needs to be made up in two pieces from one eighth steel bar, as shown in the picture. Fit diagonally within the door, on the corner of the door that needs to be pulled into the body. Rebate any wood in the door where the straightener is going to run. By tightening the nut and bolt that joins the two pieces of the straightener, the door will twist and reverse the warp.

*Straightener fitted diagonally across the inside of a door, where the trailing edge had bowed outwards. By tightening the central nut & bolt, the corner of the door is pulled back into alignment within the door frame.*

Some Rolls-Royce tourers feature polished aluminium panels. All cars feature the iconic Grecian inspired radiator, finished in either Nickel or Staybright.

In the case of aluminium, if the metal is oxidised or corroded, then you will first need to rub it down to smooth, matt metal. Begin by scrubbing the panel clean, using cellulose thinners. Then rub down using wet-n-dry aluminium oxide abrasive paper. Dip the paper frequently into water, to wash away residue. If you're working on a flat panel, then wrap the paper around a cork block. Depending on the level of corrosion, choose an abrasive grade in the 600 800 1200 2000 range. On most cars it will suffice to start with 800 and work through to 2000. When you have finished, you will be faced with a smooth satiny sheen. Now you can start to achieve a mirror finish.

The best polish, in my experience, is the Nu Shine range. But it is not cheap! This product will not only outshine other products, but it also imparts a protective film onto the metal. Nu Shine is in fact specified by both Airbus and Boeing, to be used on the leading edges of all aluminium wing panels. Consequently in the UK, you need to shop at an aviation supplier, such as Fraser's Aerospace. Use Nu Shine in the same way as you would when compounding paint. Less is more! Start with grade 'C' and the aluminium will quickly begin to shine. Then finish with grade 'S' to achieve a

mirror finish. Radiators: Rolls-Royce cars produced prior to around 1930 had German Silver and nickel-plated radiators. These cars were produced before Rolls-Royce used Staybright, an untarnishable finish that requires very little maintenance. Nickel, however, does not weather well and tarnishes when exposed to the air. It also water spots and the radiator finish will become pitted over time, if not regularly maintained. Especially in regions that experience acid rain. Nickel plating, German Silver & Staybright have a harder surface than aluminum and do not require as much polishing to achieve a good finish. If there is fine scratching, or pitting evident, then start with Nu Shine 'C'. Most radiators will come up gleaming, using just Nu Shine 'S'.

Finally, you will see on the polishing web sites, that they all recommend a special (expensive) cloth for the final polish. This cloth is flannel, which can be bought from any material shop for around £4 metre.

GXO71 as acquired was cellulose painted. But I knew from the 1930 photograph that Tom Clarke unearthed, that the car when new, was 50% polished aluminium. It was also the case that the cellulose re-paint that had been done in the USA in the 60s, had now degraded. The finish was beginning to micro blister. I decided to remove the paint from the previously polished areas and to repaint the rest, at a later date.

The car was taken to Orchard Restorations in Horam, Sussex and all four wings and the front valance were removed, plus the fillets between the bonnet and the chassis rails. All the paint was hand stripped, when it was discovered that the edge of both front wings, from the running board junction to eight o clock on the road wheel, was full of filler. When the filler was removed, it revealed that this area of the front wings resembled lace. Long strips of new wheel formed aluminium were TIG welded in place. Aluminium solder was run into the imperfections and the repairs were then planished. Similarly there were a few bad areas on the rear wings, including an old repair patch. All of this was cut out and

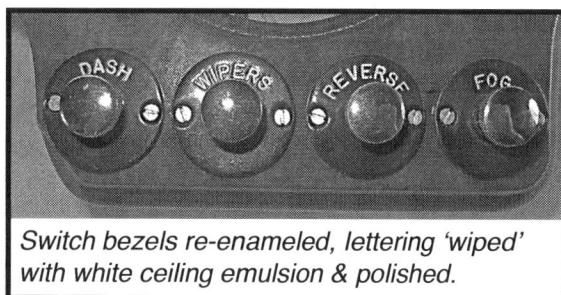

*Switch bezels re-enameled, lettering 'wiped' with white ceiling emulsion & polished.*

new aluminium let in. The old repair patch was where a support had been fixed, for the rear boot extension that the second owner built in the early '40s. While the front wings were off, the opportunity was taken to renew the headlamp wiring. After re-assembly, everything was polished by Orchard, using abrasive papers and finishing up with 1200 grit applied wet, by hand. The final result was a sheen, rather than a perfect high

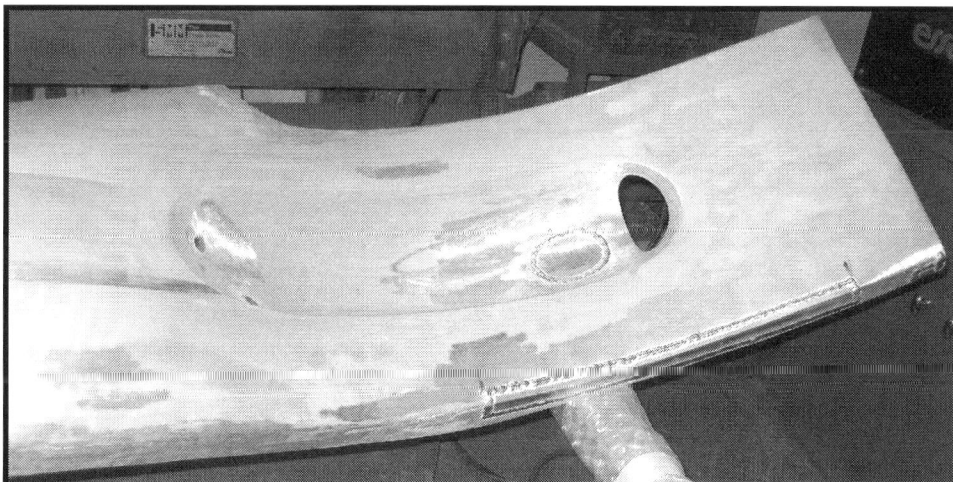

*New metal Tig welded into a front wing. It has yet to be planished, aluminium solder loaded, rubbed down and finally polished.*

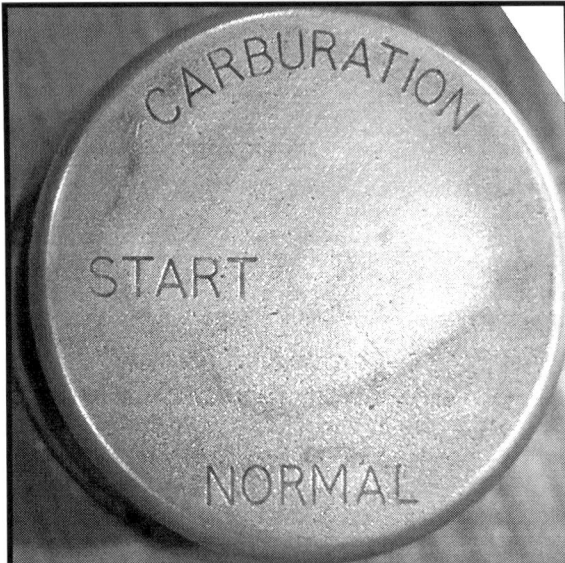

*All bezels need to be stripped of paint. A High Street engraver can then add depth to the lettering, before re-finishing as described.*

gloss shine. Total cost inc. VAT was £3,143.75.

If your car has a walnut dash and door capping's; consider re-furbishing the wood yourself. If you are not confident of stripping out the dash yourself, by all means get a professional to do it. But instead of sending the wood to an auto restorer, go and see a local antique furniture restorer. Don't use the word 'Rolls-Royce' but rather 'old car'! I used a local antique shop to do a

25/30 and got all the wood (9 pieces) re-lacquered for £350. Similarly, if you have 'tired' push/pull dash switches and starting carburettor bezel, renovate yourself. The brass switch bezels were originally stove enameled black and then engraved. Consequently when the old stove enameling is removed, there is not enough depth left in the lettering to accept white paint. I went to a High Street jeweler who offered engraving services. They refaced the bezels and then machine engraved the lettering in the style of the original. Total cost £60. After undercoating and spraying with a black enamel aerosol, the lettering was 'splodged' with domestic white emulsion paint and then lightly 'wiped' with a damp cloth covered finger. When the emulsion was dry, polishing with ordinary wax car polish removed all the white streaks on the black surround and the final visual appearance is very original.

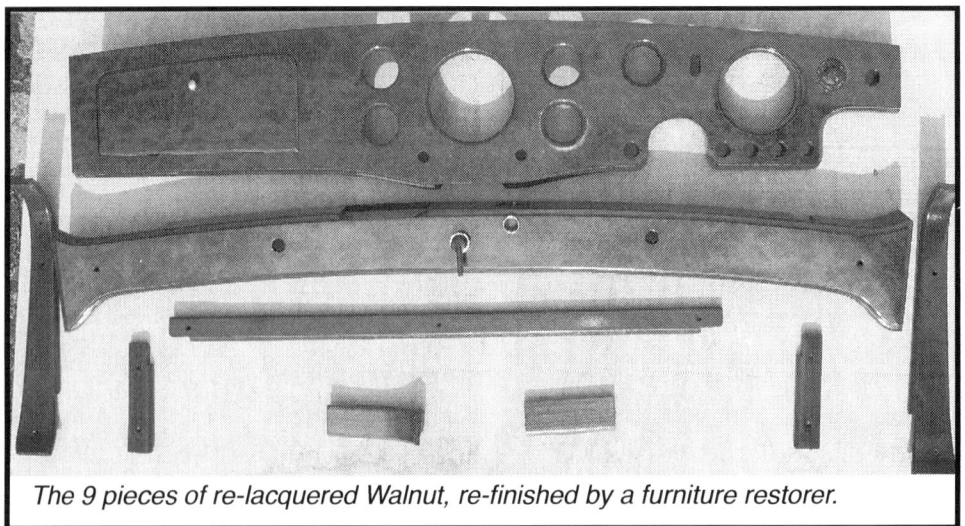

*The 9 pieces of re-lacquered Walnut, re-finished by a furniture restorer.*

## Front Valance

Henry Royce stated that the front dumb irons should always be covered. He allegedly remarked, that a lady should never reveal her underskirts. Car's that are fitted with a valance do look much better and are easier to keep clean. GXO71 was fitted with a front valance when new, but this had been discarded, together with the wheel disks, when the car was in America, in the 1960s. So I wrote to many Thrupp & Maberly 20/25 owners in the UK, enquiring about the front valance fitted to their cars and got pictures back of many different cars. From study of these pictures, I was able to establish the T&M 'house style' for the front valance. Armed with this information, a new aluminium front valance designed in the original Thrupp & Maberly style was fabricated. This was done by Orchard Restorations, Sussex, at a cost of £587.50.

*The newly fabricated front Valance. The corner's were TIG welded and then planished. Finally the piece was polished.*

## Rear mounted luggage trunk

As delivered by Thrupp & Maberly, my car had a rear mounted trunk. This was common for the period, as there is no 'boot' as such on a tourer. The original trunk had long disappeared and I attended a few automobilia auctions, in the hope of finding a replacement. But the few trunks I did locate were in bad condition. I mentioned this problem to the coach trimmer who was making new tonneau covers for the car. I discovered that he had made trunks before and didn't consider it any big deal. Working from a photograph of the original trunk, he constructed a replica. On the rear of the original trunk, there was what looked to be a logo. 'RR' within a circle was jig-sawed from thin plywood and glued on the back of the trunk. When the finished trunk was ready for covering, leather cloth was stretched over the top of the logo and beaten into place. The finished result looks just like an original embossing. Cost including all materials, was £1,000. Rather than pay the exorbitant price demanded for 'Brooks' style trunk catches, I bought four vintage bonnet catches from 'The Complete Automobilist'. They secure well and look authentic, fitted upside down on the trunk.

A previous American owner of Bluebell, wrote to tell me that he'd found the original driver's door side screen in the back of his garage. He kindly mailed me this screen. It turned out to be the Celluloid part of the side screen; the original metal frame that holds it, I discovered behind the back seat. This Celluloid part has a split hinge, to allow the driver to make hand signals when the screen is in place. The Celluloid was yellowed and cracked and the screen was beyond salvation. However, it served as an accurate pattern to remake a new one in the same style. I had the frames shot blasted and then flame sprayed with hot zinc at a cost of £60. I did not paint these frames, as the paint would chip due to the tight storage space behind the rear seat. The matt hot zinc finish actually looks quite good. New leatherette, sheets of Perspex and pop-the-dot fasteners etc. was ordered from Woolies at a cost of £165.59. Strips of this leatherette (grained vinyl) are held between the inner and outer frame at the foot and back edge. They serve to channel rain from the Celluloid window onto the outside of the car.

My wife machined the weather sealing strips on her Singer sewing machine, using the original screens as a pattern. The screw threads on the frames were also in a sorry condition

*Driver's side screen in place, showing the leatherette weather-proofing strips. Note also the triangular perspex bottom half edged in vinyl. The vinyl is sandwiched by the corner to corner piece and again held by countersunk screws. This long vinyl edge acts as a hinge, enabling the flap to rise, so that the driver can make hand signals.*

and all were re-tapped to take an M5 stainless steel countersunk screw. The two male 'fingers' on the inner frame, locate into nickel finished holes on the top of the doors. The finished result is six serviceable screens, bringing the car back to its original 'all weather' capability.

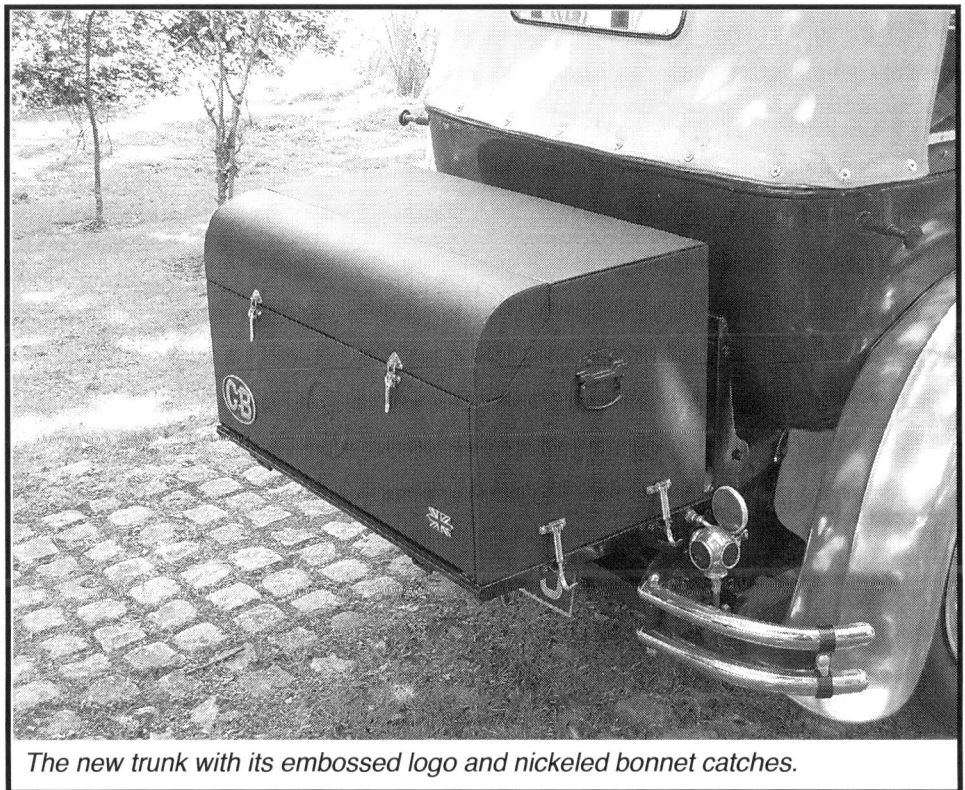

*The new trunk with its embossed logo and nickeled bonnet catches.*

A huge number of articles have been written on the subject of lubrication for the older car. I'd like to think that I've read my share. The distillation of which, has lead me to my own conclusions as to which is the 'best' oil to use. For the Rolls-Royce 20 or 20/25 engine that has not had an after market engine oil filter fitted, my oil of choice is Castrol or Halfords Classic 20/50 multigrade. If the engine has had a spin-on modern oil filter fitted, then my oil of choice is Rock Viscol 20/50 semi-synthetic.

In the gearbox, it is sensible to take advantage of modern extreme pressure oils. Use an EP80 or 90 to GL3 or GL4 specification - *not GL5*. These EP oils can withstand far greater point loadings than a straight mineral oil and are also much

'stickier'. Which means they stay in the bearing and do not drain away over time, like an ordinary oil. For the same reasons, also use GL3 or GL4 in the rear differential.

If you have a noisy gearbox, then eventually it will need to be re-conditioned. However a worn bearing in a gearbox, will probably go on for many more thousands of miles of (noisy) service. To make a gearbox quieter and improve the gear change at the same time, try mixing half a pint of Lucas Heavy Duty Oil Stabiliser with the gearbox oil. Contact details for both supply and more information, are at the back of this book. This oil additive can also be usefully added to the engine oil, if it is intended to leave the car unused for a long period of time, as it also serves as an inhibiter.

For the steering box, you need to use an oil with a moly supplement added - especially important if your car is prone to 'sticky steering'. Use Penrite Steering Box Lube. This is almost a fluid grease and you will find it easier to pour, if it is warm. As a specialist oil, this is not stocked in the usual outlets. However, it is available mail order, from the RREC Club Shop.

For the chassis lube system, ignore the Handbook instruction to use 'engine oil'. Use EP90. For the starter motor switch and for use in the oil can to the many oiling points indicated in the Service Schedule - use a straight mineral SAE 30 or 40 oil.

*The engine sump drain plug has been scribed on the face, to make it easier to insert locking wire. The scribed marks register with the locking wire holes, which you cannot see, when the drain plug is tightened.*

## Engine

1. Change oil: 10 pints Castrol Classic 20/50. Clean gauze oil strainer in sump.
2. Oil can to starter motor bearing.
3. Remove starter end cover & clean out dust.
4. Clean & gap contact breaker: .017" - .20".
5. Clean & gap spark plugs: .020" - .025". Renew every 10,000 miles.
6. Oil can to distributor advance/retard.
7. Grease contact breaker cam.
8. Fill water pump gland with red WP grease.
9. Adjust fan belt - 3/4" free play.
10. Oil can to fan pulley.
11. Blow out any debris from radiator core.
12. Drain, flush & refill radiator with 15 litres 50% anti-freeze & 50% softened water. Ensure anti-freeze has IAT & *not* OAT inhibitors.
13. Clean Autovac fuel filter on inlet pipe.
14. Clean carburettor fuel bowl
15. Clean carburettor air valve & chamber.
16. Check 12 tappet clearances: all .004" engine cold.
17. Clean breather pipe gauze crankcase to carb.
18. Oil can to 2 dynamo bearings (1 under the cover) - unless using sealed-for-life bearings.
19. Clean out brush dust in dynamo.
20. Oil can to dynamo drive coupling.
21. Oil can *sparingly* to 2 oilers on Magneto.
22. Oil gun (EP 80/90) to front engine bearer. Screw down until oil exudes.

## Under-bonnet

23. Refill Luvax chassis oil pump mineral EP 80/90.
24. Check oil level in starter switch SAE 40.
25. Oil can to Control Rod bearings on steering box, steering wheel, carb, instrument board, radiator shutters, early/late control.
26. Oil can to Klaxon Horn.
27. Oil can to bonnet & ignition lock.
28. Oil can to 4 bonnet catches.
29. Clean or replace the fuel filter element, located under the steering column.
30. Clean engine bay including inside bonnet.

31. Renew felt air bleed filter in Autovac.

## Wheels

32. Remove & grease splines.
33. Remove disks (if fitted) and check security of ring screws.
34. Check tyre pressures (600x19 6 ply) 35 lbs all wheels including spare(s).

## Transmission

35. Change gearbox oil - 2 pints EP 80/90 to GL3 or GL4. Fit new sealing washer.
36. Remove clutch pit cover. Insert two drops of engine oil into oil hole in withdrawal bearing.
37. Change overdrive (if fitted) ATM fluid - 1.25 pints (.75 litre)
38. Oil gun to nipples on prop shaft.
39. Change differential oil: 2 pints EP 80/90 to GL3 or GL4. Fit new sealing washer.

## Chassis

40. Check fluid level in shock absorbers every 10,000 miles - Castrol 'F'. (SAE 20 Mineral)
41. Check tightness of steering arms on stub axles.
42. Top up Steering Box: Penrite steering box lube.
43. Check battery levels and grease posts (96 amp Negative Earth).
44. Oil gun to O/S King Pin filler - 3 to 4 strokes over 3 to 4 minutes.
45. Oil gun to rear axle filler on differential case - 3 to 4 strokes over 3 to 4 minutes.
46. Oil gun (liberally) to all road springs (nipples are on leather gaiters).
47. Grease 2 nipples on rear shackles.
48. Adjust brakes if necessary.
49. Release *(not remove)* drain plug on petrol tank, to drain any water/sediment.
50. Check clutch pedal free play.
51. **Oil can to:**
    8 brake jaws rear axle.
    12 jaws on front/rear brake ropes.
    4 ball joints on front brake pull rods.
    (Underneath leather gaiters)

4 jaws brake rod Servo.

4 coupling joints on Servo rods.

2 joints cross shaft on Servo.

2 jaws brake pedal to cross shaft.

2 jaws clutch pedal connection.

1 throttle pedal bearing.

2 jaws hand-brake to equaliser.

4 hand-brake pawl connections.

3 reverse catch of gear lever.

3 steering wheel boss controls.

## Road Test

**52.**  Check operation of Magneto.

**53.**  Check all lights, horn, indicators.

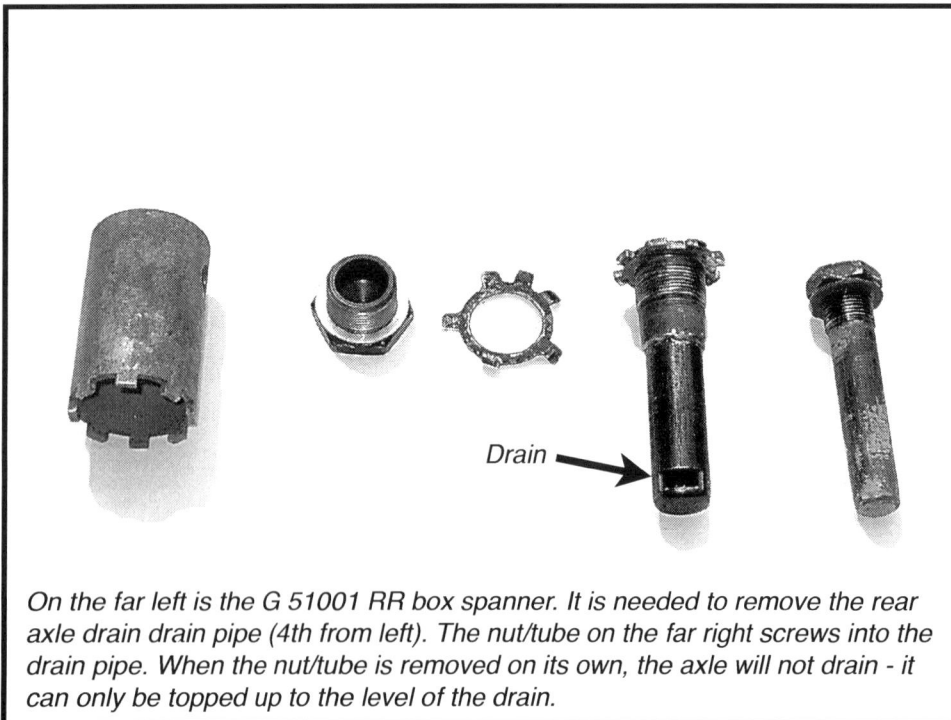

*Drain*

*On the far left is the G 51001 RR box spanner. It is needed to remove the rear axle drain drain pipe (4th from left). The nut/tube on the far right screws into the drain pipe. When the nut/tube is removed on its own, the axle will not drain - it can only be topped up to the level of the drain.*

Caveat Emptor applies in spades when contemplating buying a Rolls-Royce car. Bear in mind that these cars were hand crafted with little regard to either time or cost. It is so easy to be seduced by a car and then to spend years while it drains money out of your bank account. The best buying advice, is to try and find an existing 20/25 owner who is prepared to look at and more importantly, drive the car you have set your sights on. Ideally that owner will have some mechanical experience. But if he hasn't, it's still a worthwhile exercise, as all owners will be able to compare their own car, with it's quirks and known faults, to your intended purchase. The first decision you need to make, concerns the body style. Firstly the practical considerations. If you are tall, then a limousine with division is never going to prove comfortable. Coach builders put all the

space and comfort in the rear; most of these cars would have been driven by chauffeurs and little regard was shown for the comfort of the driver. It was 'Mr Big' in the back who had to be 'sold'. Then there are the financial considerations. If you are going to spend money on the car to bring it up to concours or near concours standard, then it makes sense to start with a tourer or a sporting saloon. These are nearly always going to appreciate more than a run of the mill 'D back'. Try to get hold of a copy of Lawrence Dalton's 'Those elegant Rolls-Royce'. That's the bible as far as coach built cars go and you can browse all the different

*GWL18: 20HP Horsefield Tourer, parked next to GXO71: 20/25 Thrupp & Maberly Phaeton Tourer. Both cars were on tour in Northern Spain.*

designs that were produced. Having decided on the body style, then the next decision is where to buy. The ideal, is privately from an owner who has cherished the car for years. Bear in mind that most of these cars are sold word of mouth and do not always get advertised. So join the appropriate geographical section of the RREC and ask around. If you spot a desirable car with an elderly owner (and there are many) make yourself known and ask if you can have first refusal, if the owner ever decides to sell.

When inspecting a car, the first thing to do is to depress the chassis lubrication pedal down to its full extent. This pedal is found on the bulkhead, to the left of the centre line. The pedal should stay down and take some minutes to return to fully up. When it does, depress it again. This action will pressurise the oil pipelines and by the time you have finished looking at the car, should result in visible oil on the garage floor. Refer to the diagram in this book, which will show the location of the 'drip valves' and will give you an idea of where to look. If there is no visible oil anywhere, then walk away. A few drip valves blocked is more likely. To free & clean them is probably within the scope of most owners. As much as anything else, the condition of the chassis one-shot lubrication system, is a good indicator of correct servicing and the general health of the chassis.

There are two main areas of potential major expense on any 20/25 - the engine and the body. All of these cars are coach built. Which means there is a body frame constructed from wood (almost always Ash) and usually panelled in metal (almost always aluminium). If the wooden

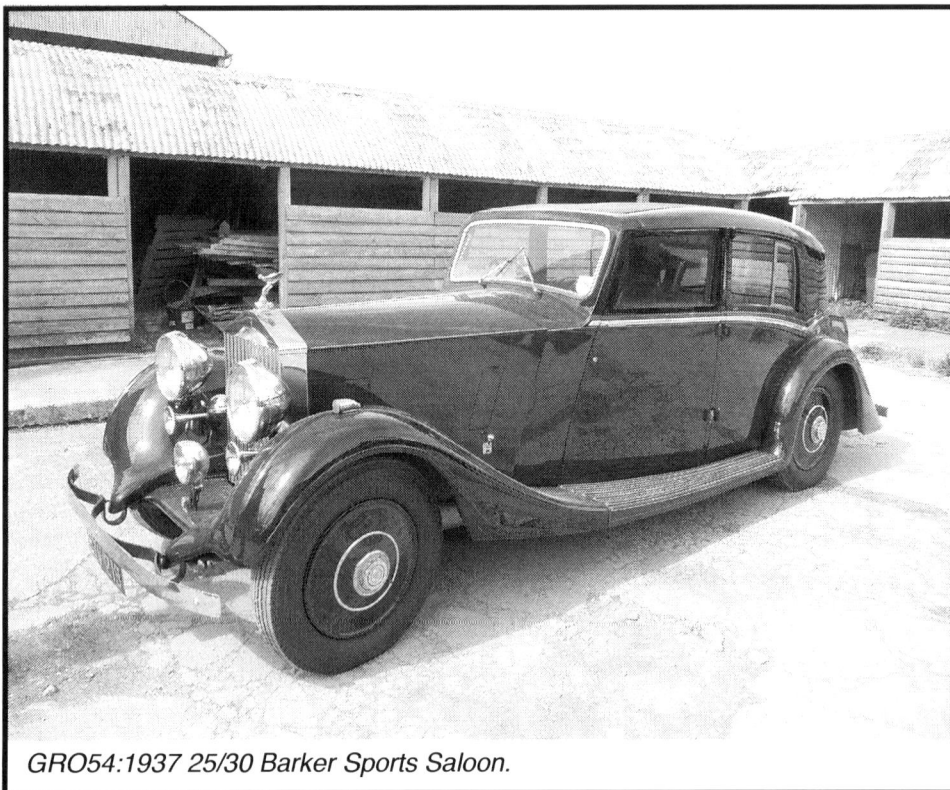

GRO54:1937 25/30 Barker Sports Saloon.

framework is rotten, then the aluminium panels have to be removed, before work can start. You could easily spend £10 - £15K and that's before repainting and which could easily double the figure. So check the body carefully. Examine all the doors. Look for up and down play when the door is open. Establish whether any play present, is in the hinges, or the door frame, or perhaps the hinge screws are fastened into rotten wood. Look for witness marks on the door aperture; has the door dropped? A good sound body will have equally spaced door gaps and all will shut with a satisfying solid 'clunk'. Open the sunshine roof if fitted and look for evidence of wood rot, water ingress, or both. Are the drain holes visible and clear? Lift the carpets inside the car and look for evidence of water ingress. Similarly, check around the windscreen; especially if it is an opening screen. Any water ingress here, will probably be revealed by water stained wood. Or worse still, veneers lifting. Finally, remove the carpets from the boot and again inspect for rot and/or water ingress.

With the car parked on a flat even surface and a driver behind the wheel, check that it is not listing due to tired springs or a broken leaf. Measure from the bottom of the front wings to the ground and do the same at the rear wings. The measurements side to side, should be within .25" of each other. I'm taking it as read, that you will also examine the body externally for cracks, parking swipes and the condition of the paintwork/brightwork generally. Also the condition of the tyres; look particularly for signs of cracking in the sidewalls. All tyre manufacturers recommend that tyres should be discarded irrespective of mileage, when they are more than 7 years old. Currently tyres for the 20/25 are £200 each, plus

GXL39: A much toured 20HP Hooper Saloon, seen here crossing The Loire.

another £25 for an inner tube, rim band and fitting.

Moving on to the engine; remove the radiator cap and inspect the coolant. It should be visually clean and filled with the correct IAT anti-freeze. This is usually coloured blue. Modern extended life (OAT) anti-freeze is likely over time to destroy the gaskets in the engine with disastrous results. So quiz the owner on this issue. Open the offside bonnet. The engine should present clean with no significant oil leaks. All parts should also be black enamelled. Whatever electrical cabling is visible, it should be cotton braided and not modern plastic cable. Similarly, the ignition coil should be the correct Rolls-Royce brown 'mushroom' type and not a modern coil. This visual inspection is to establish originality and is a good indicator that the car has been loved and properly maintained.

Start the engine and allow it to warm up. Early cars had manually operated radiator shutters and no calorstat. Adjust the shutters to show 70 - 80 degrees on the dashboard gauge. On later cars fitted with automatic shutters, check that the temperature stays within that range. When hot, the engine should idle quietly and smoothly, with circa 10lbs oil pressure. At hot idle, move the right hand lever on the steering hub (early/late) to the centre of its range. Then push the mixture control on the steering hub fully to the left, pause briefly and then push it fully to the right. In both extreme positions, the engine should slow, run roughly and perhaps even stall. It should sound sweetest, with the mixture control approximately in the centre of its range. Stop the engine. Examine the underneath of the engine for coolant and/or oil leaks. Open the nearside bonnet panel

Left: GBA64: a concours condition 20/25 Freestone & Webb Sedanca. Right: GDK22: a much earlier 20HP Hooper Doctor's Coupé.

124

and ask the owner to engage the magneto; full instructions are in the hand book. Start the engine again. It should start easily and run much the same as it did on the coil.

Take the car for an extended road test; ideally at least 20 miles. The engine should pull strongly in all gears, with no knocks or clatter. Engine vibration and/or noise, could indicate problems with the vibration damper; sometimes called the 'Slipper Drive'. Similar to the crankshaft, this damper fitted in the front wheel case can get totally seized with sludge and is a very expensive item to overhaul. When these engines left the factory, they were fitted with a very crude oil filter. It would stop a nut&bolt from circulating in the engine, but little else. Consequently if engine oil change requirements have not been meticulously followed, carbon and combustion sludge will form. The crankshaft is fitted with centrifugal extractors. But as soon as these are filled, sludge will build and especially in the vibration damper, where centrifugal

force will collect and solidify it.

Early cars will have no synchromesh and some graunching of gears is to be expected, unless you are very experienced. Later cars should exhibit a sweet gearbox, with synchromesh on third and top gear. On all cars first gear is very low and is rarely used once the car is actually moving. It is entirely possible never to use first gear, though this is not good practice on the early cars. These were fitted with a Rolls-Royce clutch and it will not stand a lot of abuse. This clutch should be viewed more as an in-out device and should not be slipped. Later cars had Borg & Beck clutches and these were more robust to suit the average owner driver. All cars are very low geared, as Rolls-Royce had to cater for a wide range of body weight. Consequently

49G1: A stylish 20HP Park Ward Doctor's Coupé.

all will benefit from the fitting of a Laycock overdrive. This overdrive drops the top gear revs by 25-28% and allows the car to cruise all day unstressed at 50-55mph. Some steering shake when driving across a pot hole is to be expected. The brakes are powerful and braking hard should pull the car up in a straight line. Even the early cars should have no problem in coping with modern traffic conditions.

If the car is fitted with wheel disks, remove them and examine the wheels. Wheel disks sometimes hide broken and/or missing spokes. Or visible rusting of the rims and spokes. Wheels in this condition will need rebuilding and together with painting will set you back in the region of £250-£300 each. With the car jacked up at the front, a procedure which will also reveal the condition and presence of the jack and large tools, rock the road wheels in turn and feel for play in the suspension. Spin the wheel and visually check that it runs true with no noise. Then remove the wheel and visually check the condition of the splines on the wheel and the hub. Fresh grease should be evident on the splines and the spline serrations clean cut with no burring or damage.

Ideally, following a road test, conduct a compression test on the engine, to establish that all six compressions are roughly equal and in the range 90-120lbs.

All of the above assumes that you have already grilled the owner regarding maintenance and have examined the paperwork relating to this. However low the annual mileage, these cars should be serviced at least once a year. So if the maintenance history is sketchy or absent, walk away from the car, or adjust the price offered for the car accordingly.

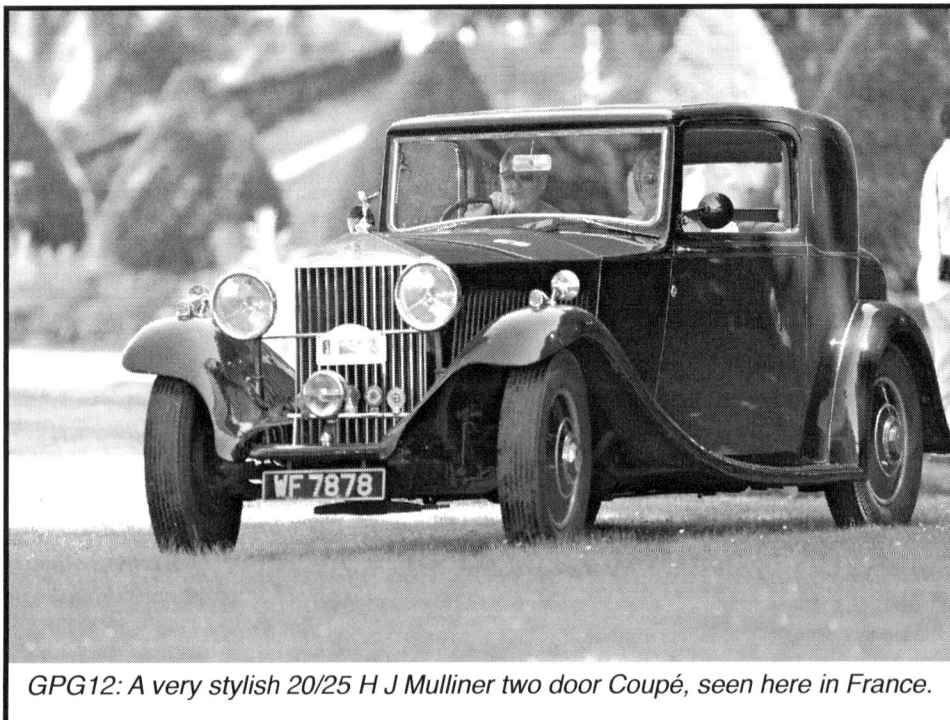

*GPG12: A very stylish 20/25 H J Mulliner two door Coupé, seen here in France.*

## Matching Numbers

The value of any Classic or Vintage car is enhanced if the car has matching numbers. For the majority of makes, this means a chassis and an engine number on the car, that matches those numbers recorded in the Registration Document and/or Log Book. Rolls-Royce went further than this. On all pre-war cars, the Derby factory recorded the number of almost every discrete component on the chassis. Which means that it is possible to check on any given car, not just that the engine is original, but also the magneto, the carburettor, etc etc. A copy of these Chassis Cards, plus the original specification, sales order and chassis test data are available for every Rolls-Royce chassis. It is a major benefit of joining the RREC.

## Starting the engine (early cars)

1.    If the engine has not run for a very long time (months) then it is advisable to turn the engine briskly on the handle with the ignition off, until oil pressure registers on the gauge. To do this easily, remove the sparking plugs.

Open the offside bonnet panel. Lift against the spring and rotate the catch that secures the top of the carburettor float bowl. Unscrew the float bowl cover about four turns. Wait a few seconds until fuel gushes out of the central overflow and then quickly screw the cover back home. Re-engage the securing catch.

*Do not completely unscrew and release the float bowl cover, as this will*

*GHJ50: A lovely 20HP Park Ward Drop Head Coupé suitable for Bertie Wooster.*

*result in an uncontrollable flood of fuel,*
*until the Autovac reservoir has emptied.*

**2.** For a normal cold start; turn on the Petrol Tap which is located inside the car, in the front passenger footwell. Turn the lever upwards for 'ON'.

**3.** Turn on the Battery Master Switch (if fitted).

**4.** Turn the shutters control to fully closed; all the way to the left.

**5.** Set the early/late ignition control to its mid position. This is the right hand knob on the steering wheel boss.

**6.** Set the mixture control Rich, by swinging the lever three quarters to the right. This is the small brass lever at 12 o'clock on the steering wheel boss.

**7.** Set the throttle idle control to its mid position. This is the left hand knob on the steering wheel boss.

**8.** Set the starting carburettor to ON. This is a lever on the dashboard.

**9.** Put your hand on the ignition switch, but do not operate it yet. This is the right hand brass lever on the black circular switch box above your right knee.

**10.** Check the gearbox is in Neutral by observing the position of the gear lever in the gate.

**11.** Press the starter pedal, located just to the left where your left foot normally rests. As soon as you hear the starter operate, flick the ignition switch one position UP. Engine starts; release the starter pedal.

**12.** As the engine warms up, quickly but progressively, turn the starting carburettor

lever to OFF. At the same time, increase the throttle 'idle' to maintain revs. Open the radiator shutters as the engine warms up to maintain 70-80 degrees. As you drive off, move the early/late control all the way up, to full early and bring the mixture control back to the centre.

**13.** Starting when the engine is warm - set the throttle idle three quarters of the way up the steering wheel quadrant and then follow instruction 11.

### The correct use of the early/late control

The steering wheel early/late (advance & retard) control, is used to alter the point at which the spark plug fires. This is not, as many owner's think, when the piston after drawing in gas, is at the top of its stroke. Although the engine will run, if the spark occurs at TDC (Top Dead Centre), it will not operate efficiently. This is because it takes time after the spark has occurred, for the flame to become established and to burn all of the gas mixture. Therefore the spark is timed to occur, as the piston is still traveling up the bore, so that as the piston reaches TDC, flame propagation is peaking and all of the mixture is burnt.

However, if the engine is idling very slowly, the pistons are not moving fast enough, to reach TDC at the same time as maximum burn. So moving the ignition control down towards the late position, will fire the sparking plugs when the pistons are closer to the TDC position. The consequence of this, is that the engine will idle slower and more smoothly. Similarly, if you attempt to start the engine on the handle, with the control in the fully early position, then the gas in the bore is likely to ignite and exert more force on the rising piston, than the energy you have inputted with the handle. The result, will be that the rising piston will be stopped dead and then forced backwards. A so called kick-back, where the force will be transmitted violently to the starting handle and may well injure your hand. So when starting the engine by hand, it is essential to move the control down and good practice to do this, even when starting using the self starter. At all other times, the ignition control should be set at fully early, which will extract the maximum power from the burning mixture. All of this pre-supposes, that the engine has been correctly set up in the first place.

***Principal changes; early cars.***
GXO 11-111
1929/30
Wing mounts for headlamps. Double arm spare wheel mount.
GGP 1-81
1930
GDP 1-81
1930
GWP 1-41
1930
GLR 1-82
1930
3" longer chassis to 132". Bigger throat on carburettor. 5.25:1 compression ratio.

GSR 1-81
1930
GTR 1-41
GNS 1-81
1930/31
Grooved leaf road springs.
GOS 1-81
1931
GPS 1-41
1931
Revised top to Autovac. Headlamp pillars mounted on chassis. All black steering wheel.
GFT 1-81
1931
6 vane water pump. 2 filters in petrol tank. Hinged petrol filler cap.
GBT 1-82
1931/32
Thermostatic shutters from GBT22. MAI flywheel mark.
GKT 1-41
1932
Synchromesh gearbox from GKT 22.
GAU 1-81
1932
Electric petrol gauge
GMU 1-81
1932

*Rolls-Royce cars of this era were subject to continual development and consequently there can be significant differences between different 20/25 chassis'. The specification below is correct for GXO71 This first series 'O' range, ran from GXO11 to GXO111. (1929-1930)*

**Engine:** Six cylinders, monobloc with detachable head. 3.25 inch bore x 4.50 inch stroke. 3,669cc (25.3hp RAC). Unit construction with gearbox mounted in a sub-frame attached to the chassis at three points. Compression ratio 5.25 to 1 (Originally was 4.6 to 1; most engines have now been rebuilt with the later compression ratio). Firing order 142635. Oil Pressure (minimums): 4 lbs at idle. 25 lbs running (hot).

**Engine oil:** Full capacity is 10 pints.

**Spark Plugs:** NGK BP6ES (plus thread adaptor).

**Valves:** Overhead, operated by pushrods. Tappet clearance cold 0.004" inlet & ex.

**Crankshaft:** Carried in seven bearings with slipper drive vibration damper at front end.

**Camshaft:** Carried in seven bearings.

**Lubrication:** Gear-type pump delivers oil at full pressure to crankshaft bearings, big ends and gudgeon (wrist) pins. Relief valve lowers pressure for timing wheel case & valve gear.

**Electrical:** Independent coil and stand-by magneto systems, used separately. Advance & retard control on steering wheel. 12 volt system with a 50 a/hr (originally) wet battery. Dynamo charge controlled by driver. Contact breaker gap 0.017 - 0.021".

Magneto points gap 0.014". Sparking plug gap 0.020 - 0.025".

**Cooling System:** By centrifugal pump and belt driven fan. Driver operated radiator shutters with warning light on dash to indicate 90°C. Water capacity 30 pints.

**Carburettor:** Early cars 'N series' - two jet type with steering column mixture control and automatic air valve. Hand throttle on steering wheel.

**Fuel System:** Autovac fuel pump powered by manifold vacuum. 14 gallon fuel tank mounted at rear of chassis. On/Off fuel tap mounted on inner bulkhead (Separate on/off & reserve fuel tap retrospectively fitted on top of fuel tank). Hobson hydrostatic fuel gauge mounted on dash (early cars).

**Transmission:** 4 speed gearbox. Gear ratios: 1st 3.73:1 - 2nd 2.33:1 - 3rd 1.49:1 - 4th 1:1

**Oil capacity** 2 pints.

**Overdrive:** Laycock 'J type' 28% overdrive retrospectively fitted. 1.25 pints ATM fluid.

**Clutch:** Single Rolls-Royce dry-plate type. Propellor shaft: Open shaft with enclosed oil-retaining universal joints.

**Final Drive:** Spiral bevel 4.55:1 drive with fully floating axle. Oil capacity 2 pints.

**Brakes:** Internal expanding, mechanically servo assisted on all 4 wheels. Independent handbrake operating on rear wheels.

**Chassis Lubrication:** Partial centralised pump system operated by driver. Plus oil gun to a total of 16 nipples.

**Suspension:** Semi-elliptic springs, front & rear. Hydraulic shock absorbers.

**Steering:** Worm & nut.

**Wheels:** Well-base wire 'Colonial' type with 3 sets of radial spokes per wheel. Tyres 600x19 normal inflation 35lb/in2

**Chassis Details:** Wheelbase 129 inches. Track 56 inches. Length: 15' 3" not including trunk. Width: 5' 10". Ground clearance: 8".

**Turning Circle:** RH 47.5 feet. LH 42 feet.

**Weight:** Chassis complete with tyres, battery, petrol, oil & water but excluding lamps - 2,904 lbs
Body - 784 lbs
Car complete 3688 lbs

**Handbook:** Version IX

---

## Tools as originally supplied

*(Source RR General Arrangement List F54927 1929 -1932). All tools BSF or BA.*

JAW SPANNERS single ended:
F 51960 - 5BA
F 51961 - 3 BA
F 51962 - 2BA
F 51963 - 1BA
F 51964 1/4" - 2 off
F 51933 - 5/16"
F 51934 - 3/8"
F 51936 - 1/2"
F 51935 - 7/16"
F 52298 - 11/16"

BOX SPANNERS (all 4.5" long):
F 9840 - 7BA X 5BA

F 9809 - 3BA X 2BA
F 9810 - 1BA X 1/4"
F 9811 - 5/16 X 3/8"
F 9812 - 7/16 x 1/2"
F 9814 - 3/4 X 13/16"
F 54932 - 5/8"

F 52716 - Pivot Nuts
G 51004A - Rear Axle tube nut
E 53792 - Sparking Plugs

TOMMY BARS:
E 7342 - 3/16"
E 7658 - 1/4"
E 9048 - 5/16"
E 12645 - 7/16"

G 50976 - Rear Axle tube nut
RING SPANNER:
E 52018 - To remove carb filter
C SPANNERS:
D 51637 - Starter Motor end bearing nut
F 51789 - Steering column universal joint nut
E 55017 - Water Connection Nut
E 6258 - Gearbox tower nut & Exhaust pipe
F 51754 - Rear shock absorber casing
E 6422 - Prop Shaft & Steering Column

CASTELLATED BOX SPANNERS:
E 17488 - Camshaft Nut
G 51001 - serrated box; oil drain differential
BOLTS/RODS:
G 3425B - Hub Withdrawal screw with collar
G 8522 - Withdrawal Rod rear axle drive coupling.
E 51653 - Dynamo Drive forcing bolt

IGNITION:

D 50837a - Distributor Spanner/feeler

E 5265A - Watford Magneto Spanner

D 50912 - File and handle for contact points

D 51968 - Carborundum paper for contact points.

T SPANNERS:

F 82810 - Petrol Tank drain

F 52219 - Cross Steering Tube adj. screw

VALVES:

E 52143 - Valve spring replacement tool

E 52142 - Valve Grinder

WHEELS & TYRES:

Dunlop tyre lever

Schrader Tyre Pressure Gauge

F 77376 - Nesthill No 9 Foot-pump

Dunlop No 10 Jack

Dunlop wheel spanner

G 52664 - hub retaining nut

VARIOUS:

RF 6613 - Lucas Girder adjustable spanner

E 52106 - Spring drive & clutch spring cap spanner.

F 81889 - 3" Screwdriver

F 81888 - 4" Screwdriver

RF 5949 - 5" Engineer's Pliers

F 77247 - Exide Hydrometer S1

1 medium rubber head mallet; black enameled metal handle.

K 210 - Bolts for withdrawing water pump gear ( 2 off).

520/3780 - 6" Gas Pliers

F 53557 - Feeler Gauges

F 77247 - Hydrometer S1

F 82659 - Nesthill oil syringe

Grease Can

No 40 Lucas oil can

Fiennes Restoration: Tel: 01367 810438.
http://www.fiennes.co.uk/Parts/Catalogue
All mechanical spares and a full restoration service.

Ristes  Tel: 0115 978 5834
http://ristesmotors.co.uk/content/
All mechanical spares and a full restoration service.

Vintage Restorations Tel: 01892 525899
http://www.mgcars.org.uk/vr/
All instrument restoration.

Rock Oil
http://www.classicrockoil.co.uk/
Semi-synthetic classic motor oil, mail order.

Penrite Steering Box Lube
http://www.penriteoil.com.au/
uk_locator.php
Specialised steering box oil.

Bryan & Son
Tel 01892 544635
Radiator re-coring & rodding.

www.hoseworld.com.
1 metre hose lengths available mail order.

Alan Murcott Tel: 01213 537775
alan@patalan.co.uk
Modern thermostat kits.

Nostalgia Cars Tel: 01823 444991
http://www.nostalgiacars.co.uk/
overdrive.html
Overdrive kits and fitting service.

Woolies Tel: 01778 347347
http://www.woolies-trim.co.uk/
All trim, carpets, upholstery etc.

http://www.poplargreg.com
Reproduction vintage tax discs.

A C Blasting Tel: 01580 201491
Grit & bead blasting. Hot zinc and aluminium spraying.

R J Lonsdale Tel: 01677470608
Stainless steel exhaust systems.

Vintage Supplies Tel: 01692 406510
http://www.vintagecarparts.co.uk
Spark plug adaptors, lighting & switches.

Wefco Tel: 01934 832027
http://www.wefco-gaiters.com
Leather gaiters for road springs.

Motor Wheel Service Tel: 01753 549 360
http://www.mwsint.com
Spoked wheel re-builds.

Frasers Aerospace Tel: 020 8597 8781
http://www.frasersaerospace.com/
Nuvite aluminium polish

Auto Electric Supplies Tel: 01584 819552
http://www.autoelectricsupplies.co.uk.
Flashing indicator components etc.

Phil Cordery Tel: 01248717808
philipcordery@btinternet.com
Fixed price complete re-wire.

Battery Brain
sales@dextera.uk.com
Key fob operated battery master switch.

Complete Classics Tel: 020-8660-9525
www.completeclassics.fsnet.co.uk/
2025.html
The 20/25 'Bible'. A must-have book by
Tom Clarke.

Colonnade Plating, Wembley, London
http://www.colonnademetal.com/
Chrome and Nickel Plating.

London Chroming Company
Old Kent Road, London. Tel 020 7639 6434
http://www.londonchroming.co.uk
Chrome and Nickel Plating.

Stafford Vehicle Components
http://www.s-v-c.co.uk
Torpedo light conversions, flashing
indicators.

http://www.vintagemotorspares.com
Armoured and cotton braided electrical
cable. Original style HT cable.

http://www.vintagecarparts.co.uk/
All manner of accessories for the vintage
car.

http://www.vehicle-wiring-products.eu/
VWP-onlinestore/sleeving/sleeving.php
Slit conduit & all cabling products.

www.sykes-pickavent.com
Compression testers.

RREC - 'The' Car Club.
http://www.rrec.org.uk/How_to_Join/
Register.php
The Rolls-Royce & Bentley Enthusiasts Car
Club. Chassis records & factory drawings
available to members, plus active social
events at Sections all over the UK.

http://www.accumate.co.uk
Battery optimisers. (trickle chargers)

http://www.lepsons.com
Gillingham, Kent
Wire wheels blasted and properly painted.

http://www.lmarr.com
All types of wheel disk & fixings. Note: this
Company is in the USA and delivery
charges are not cheap.

http://www.lucasoil.co.uk/
store_category.php?category=3
UK Mail Order web site
Lucas Heavy Duty Oil Stabiliser.

Printed in Great Britain
by Amazon.co.uk, Ltd.,
Marston Gate.